KEEPING BEES

KEEPING BEES

looking after an apiary

VIVIAN HEAD

Picture credits
All images Shutterstock except for p.35 (Omlet) and p.55 (Flow®)

This edition published in 2023 by Sirius Publishing, a division of
Arcturus Publishing Limited,
26/27 Bickels Yard, 151–153 Bermondsey Street,
London SE1 3HA

ISBN: 978-1-3988-2605-2
AD011009UK

Printed in China

CONTENTS

Introduction

Right: *You can start an apiary in almost any circumstances – as here in a large back garden, but also on a rooftop and balcony.*

Beekeeping is a fascinating hobby for people of all ages, and it is easy to become infected by what apiarists call 'bee fever'. Even if you have a fear of bees, a couple of hours in the hands of an experienced beekeeper will soon alleviate these qualms. The intricacies of the bee hive and honeycomb will soon have you enthralled and the gentle buzzing will become a pleasure rather than a dread. As with many species, the honey bee is now in decline. A downturn in our native plants means that bees are finding it increasingly difficult to locate their nests so we need to do everything we can to attract them back into our gardens. The greater the plant diversity, the more bees you will attract and support.

Unlike many other hobbies, bees do not need constant attention: they have a job to do and they will continue to do it regardless of human intervention. As long as you keep an eye on your hive to make sure the queen is reproducing, and a little more attention when you want to harvest the honey, you will probably not need to visit your hives more than once in every ten days. Over the winter period you can leave them alone completely, as they will go into a state of dormancy.

Bees do not need a large area, so even if you have a tiny garden you can still encourage them into your space. Even city dwellers can keep bees, and they are quietly buzzing in cities and on rooftops all around the globe. Before starting to keep bees, do the following:

1 Join your local beekeeping association.
2 Attend a beginner's course to see what is involved.
3 Read at least one recommended book on the subject to grasp the basics.

If this this inspires you to start keeping your own bees, you will be embarking on a hobby that is absorbing and rewarding.

The history of the honey bee

The earliest record of humans interacting with bees comes from a Spanish rock painting thought to be 6,000 to 8,000 years old. Paintings have also been found in other parts of the world and an ancient Egyptian papyrus dating back to 256 BCE records a beekeeper with 5,000 hives.

Honey was a component of over 500 Egyptian medicines, while beeswax and propolis were also used in the embalming process. Early Greeks and Romans kept bees, and Greek athletes used honey – the 'nectar of the Gods' – to boost their performance. The philosopher Pliny drank a glass of honey and cider each day to cleanse his system and promote good health.

The Bible, ancient scrolls of the Orient, the Talmud, the Torah and the Koran all mention the honey bee and the healing food that it produces. In Greek myth the nymph Melissa cared for the infant Zeus BYfeeding him nectar plundered from hives. However, while protecting the infant, Melissa was turned into an insect. Zeus turned her into a honey bee so she could make honey for eternity.

Early European monks kept bees, using their wax to make candles for the monastery. Mead, of fermented honey and water, is possibly the earliest known fermented drink of any kind and can be created naturally without human intervention. So it is possible that humans' first experience of intoxication may have sprung from the spontaneous fermentation of honey in an old tree trunk containing a bee colony.

Later, in the Middle Ages people started to cut down trees and arrange them into apiaries. A few hundred years ago beekeepers discovered that if you placed a box of straw over the top of a hive, the bees would start to store honey in it. At that time the bees would have been killed at the end of the season so that the wax and honey could be taken.

Honey bees were not found in North or South America, Australia or New Zealand until Europeans settled there, but by the 1600s, records show that the honey bee population was widespread on the east coast of America. They expanded throughout North America during the 18th century. Many Europeans who were fleeing war, poverty, strict land laws or religious persecution brought with them extensive beekeeping skills.

The 19th century saw beekeeping become commercially viable for the first time. The movable frame hive, the smoker, the comb foundation maker and the honey extractor were all invented early in the century. A fifth invention – a queen grafting tool – allowed beekeepers to control generic lines for the first time. In 1922, the USA passed the Honey Bee Restriction Act, in an effort to protect bees against the tracheal mite (*Acarapis woodi*), which attacked the respiratory system of adult honey bees. European beekeepers were losing as much as 50 to 80 per cent of their colonies and the US was anxious to avoid its introduction on home soil. The ban was partially rescinded in 2004, although movement of bee colonies was still under supervision.

Freedom of movement and attempts to mix different bee races has caused major problems with one race trying to eliminate another. The honey bee is able to adapt to minor changes in global warming, but Colony Collapse Disorder (CCD) is a bitter reminder that mankind is upsetting the balance of this delicate little worker. CCD is a phenomenon in which worker bees from a colony suddenly disappear. This first became a problem in North America towards the end of 2006, and it is economically significant because bees are needed to pollinate agriculture crops throughout the world. European beekeepers observed similar problems and it has now become a global problem. The cause of the syndrome is not yet fully understood, but they believe mites, genetically modified crops and global warming in general could all be significant factors. It is now down to us to help protect our environment by rebuilding honey bee colonies and providing them with a safe, natural habitat.

Right: *Skeps, made of straw, have been used to house bees for more than 2,000 years.*

CHAPTER ONE

UNDERSTANDING THE HONEY BEE

A single honey bee could not live for very long without the support of its colony or 'family'. A worker bee cannot reproduce, the queen is unable to produce the wax comb, collect pollen or even feed herself, and the drone's only role is to mate with the queen. That is why the honey bee family needs to work as a single unit.

The superfamily

Bees belong to the insect family Hymenoptera, a group which includes ants, wasps and sawflies. All are extremely beneficial to the environment, either as natural enemies of insect pests or as pollinators of flowering plants.

Although the actions of a colony may seem like chaos, in fact every move has a purpose and as you learn more about keeping bees you will understand that it is a highly organized society, with each bee having a clearly defined role.

Before you can fully understand the workings of your beehive, you should learn about the inhabitants and their various roles – the queen, the worker and the drone.

BEE DEVELOPMENT

Each bee starts its life as a small egg which is laid by the queen in the bottom of a wax cell built specifically for this purpose in the comb. The egg will hatch after just three days and the bee begins its larval stage inside an open cell. The larvae will be constantly fed by nursing bees first on royal jelly and then on a mixture of pollen and honey. If the egg is destined to be a queen, then it will be fed solely on royal jelly. After a further five days (six for the drone), the worker bees will cap the cell, and the larvae will start to spin a cocoon around itself. This is the start of the pupal stage, or the time when the larvae gradually changes into an adult bee. Once the bee is fully formed it will start to chew its way out of the cell cap, to emerge as an adult. The time it takes for the egg to develop into the adult bee differs considerably between each class. The queen will emerge after 14 to 17 days, the worker 16 to 24 and the drone 20 to 28 days, depending on the environment and the quality of food available.

A colony normally has a single queen, 50,000 to 60,000 workers at its peak, and several hundred drones during late spring and summer.

Right: *A queen bee emerging from the notably larger 'queen cell' in which she has completed her development.*

THE QUEEN

There is generally only one queen bee per colony and her only role is to mate and lay eggs. She is only fractionally larger than a worker and in a very busy hive she can be quite difficult to detect. This is something which becomes easier with experience, and it is a very important part of beekeeping. Her body is usually longer than either that of the worker or the drone, especially during the egg-laying period when

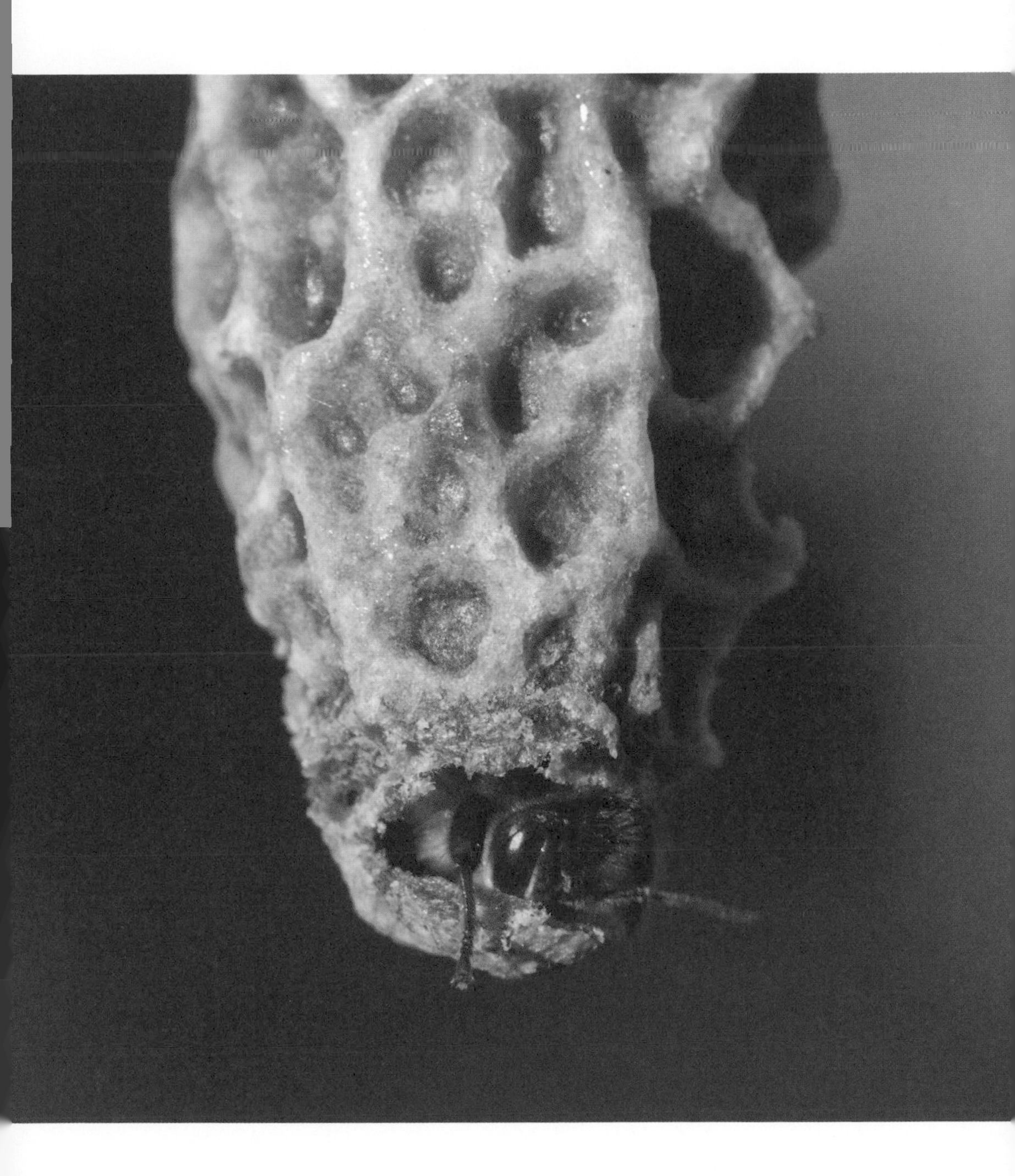

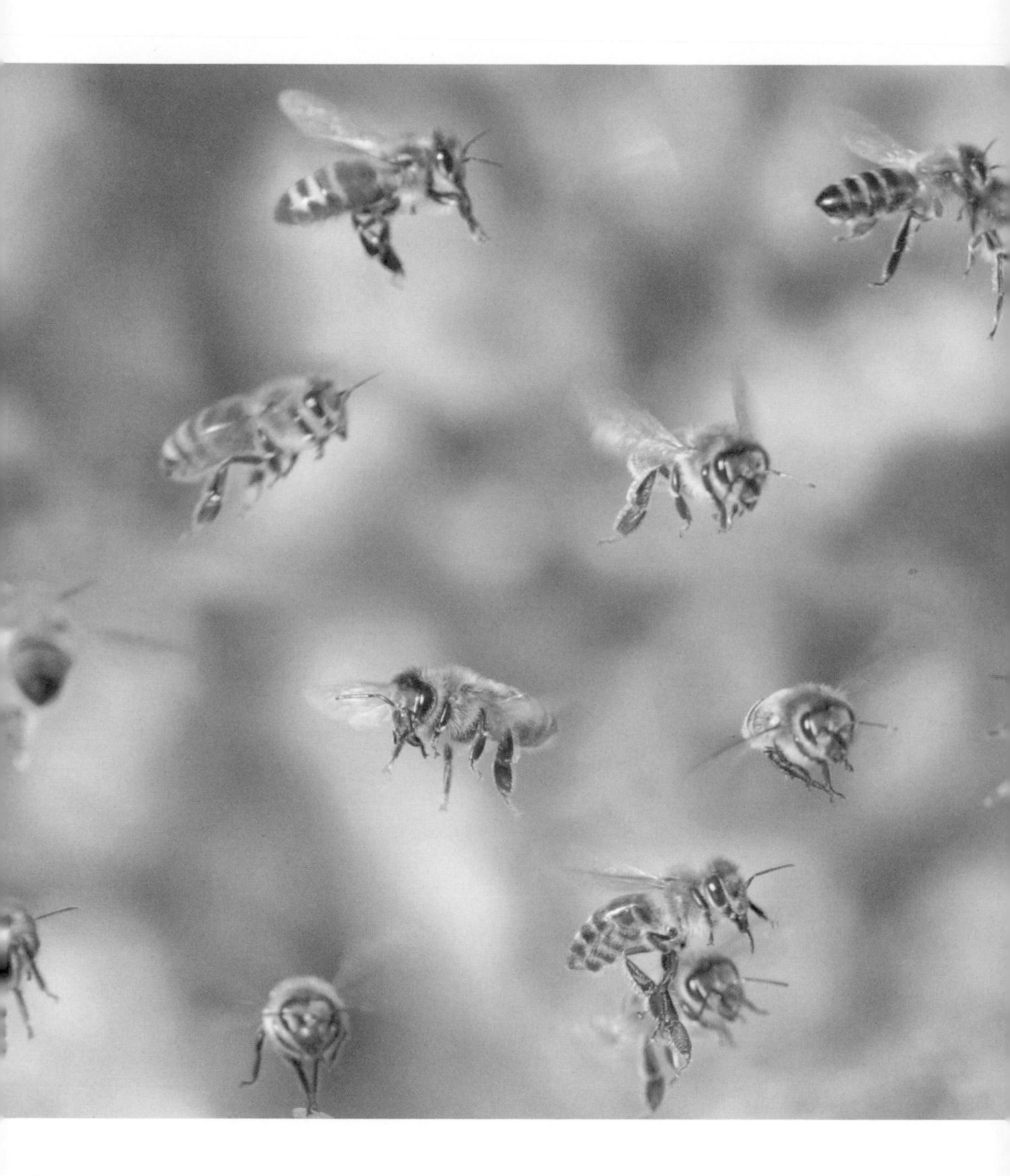

her abdomen is greatly elongated. Her wings cover only about two-thirds of her abdomen, whereas the wings of the other bees nearly reach the tip of the abdomen when folded. Her stinger is curved and longer than that of the worker and has fewer and shorter barbs.

Left: *A group of honey bees in flight. They mate on the wing when conditions are suitable.*

The mating ritual

After the virgin queen emerges from her cell, with a little encouragement from the workers and weather permitting, she will make her maiden flight within a few days. She will not leave the hive if the weather is windy or very wet. As she needs to fly some distance to locate the drone congregation area (DCA), she will first circle the hive to orient herself to its location. She leaves the hive on her own and is usually gone for around 15 minutes. She will usually mate in the afternoon and this occurs on the wing with approximately 15 to 20 drone bees. Her pheromones will only attract the drones if she is flying at an altitude above 6 metres. Each one waits its turn, then flies up to the queen and grasps her from behind before the final act of mating. As each drone completes the act, its body will literally rip apart from the effort and it dies on the spot. Then the next drone takes over and so on. Her time outside the hive is rife with danger because of predators, such as birds, and also the risk of bad weather, so the queen only makes one flight.

The queen is a vital element to the bee colony as they depend totally on her chemical production and egg laying. It is her genetic make-up, along with that of the drones she has mated with, that determine the quality, size and temperament of the colony.

Above: *A marked queen bee (centre) on the honeycomb, surrounded by a group of workers. She is in the egg-laying phase of her life.*

Laying the eggs

As soon as the queen has accumulated enough sperm in her sperm sac (spermatheca), she will return to the hive and start her life as queen of the colony. This sperm will last her for the remainder of her life, which she spends as an egg-laying machine. She starts laying within 48 hours of her return to the nest, and may lay as many as 50,000 eggs during her prime. She produces both fertilized and unfertilized eggs. Queens lay the greatest number of eggs during spring and early summer, gradually starting to slow down production in early October and do not begin laying again until January.

The queen measures the size of the cells with her antennae before laying one egg at the base of the cell. If the cell is 'worker' size, then the queen will fertilize the egg as it passes out of her. Around 21 days later, the worker bee emerges, having inherited the genes from both the father and mother. If, on the other hand, the cell is 'drone' size, the queen will not fertilize the egg and drone bee larvae will therefore form (see page 22 for more details).

The queen will be constantly attended and fed royal jelly by the worker bees. This is a vital role for the worker bee, as the number of eggs the queen lays will depend on the amount of food she receives and the size of the worker force capable of caring for her brood.

The queen bee can live for as long as five to seven years and after the first couple of years her sperm supply will start to slow down and the colony will make the decision to replace or supersede her.

Supersedure

When the queen's sperm supply begins to slow down, the workers prepare to replace her; this procedure is called supersedure. This process begins when the workers construct special cells called 'queen cups' to hold the replacement queen bee larvae. This larvae is identical to that of the worker bee at first, but the workers start to feed the larvae with a steady diet of royal jelly which allows them to mature into queens. As soon as the new queen emerges from its cell, she will immediately look for any other rival queens and kill them before they can emerge. If the old queen is still in the hive, she may kill this as well in one-to-one combat. Alternatively, the worker bees may kill the old queen themselves, surrounding her with their bodies until she overheats and dies. After the old queen has been removed from the colony, the new queen embarks on her mating flight and the whole process repeats itself.

Emergency queens

If the queen dies unexpectedly, the workers will not have time to go through the supersedure process. However, as the queen larvae is initially identical to the worker bee larvae, the workers can quickly turn this larvae into 'emergency queens' by feeding them royal jelly and making their cells larger. The first emergency queen to emerge from her cell will sting the others to death while they are still inside their cells, to ensure that she has the prominent position in the colony.

THE WORKER

The worker bee is an incomplete female that lacks the full reproductive capacity of the queen. The worker is the busiest bee in the colony, as the name suggests, and is the one you will most commonly see as they collect nectar and pollen from flowers. Worker bees pass through various task-related phases as they age.

WORKER DUTIES

When they emerge from the cell as an adult bee, the worker starts immediately on her household chores. Her six-week lifespan in summer is devoted to carrying out the many tasks necessary for colony development and survival. Many of these duties are the result of the

Right: *A worker bee gathering pollen from the stamens of a flower. The pollen sacs on her legs are already fairly full at this point.*

physiological changes that take place during the worker's life. The most important of these are the production and secretion of royal jelly and beeswax.

In addition to their numerous household duties, worker bees also forage for nectar, pollen, water and propolis. Propolis is the resinous substance collected by bees from the leaf buds and bark of trees, especially poplar and conifer trees. Bees use the propolis along with beeswax to construct their hives. Workers also serve as scouts for finding these materials and are responsible for finding new homes for a swarm. Workers are in charge of maintaining the temperature of the brood chamber, which must be kept constant at around 35°C to incubate the eggs. If it gets too hot, the worker collects water and deposits it around the hive. Then they fan the air using their wings, causing a cooling effect by evaporation. If the brood chamber becomes too cold, the workers cluster together in order to generate body heat.

The three distinct phases in the worker bee's life are as follows:

1 The 'nursing' stage lasts for about one week. At first she assists in the incubation of the new broods and in the preparation of new brood cells. Next comes the feeding of the older larvae with a mixture of honey and pollen. About three days later the special brood food glands in the head of the worker bee come active. The concentrated milky solution from these glands is called 'royal jelly' and is fed to the queen larva in its pure form, while the other worker and drone larvae are fed with a mixture of pollen, honey and royal jelly.

Below: *Worker bees tending to honey cells.*

2 Next, the young worker bee will take on the domestic phase of its life, which will last for about one week. During this phase it has various duties such as storing honey, building and repairing the comb and keeping the hive clean by removing any debris, including dead bees. It is also during this period that the young worker bee takes its first orientation flight and may also carry out guard duties at the entrance to the hive.

3 The final stage is that of the 'forager', when the bee is about 14 days old. Foraging can last for two, three or even four weeks according to the amount of energy expended on each trip. They forage for four different products – nectar, which is converted into honey; pollen, which is the protein and fat portion of the bees' diet; water; and propolis or bee glue, as it is used to close small openings in the hive. The nectar is stored in the 'crop' or honey sac where enzymes start the conversions, while pollen and propolis are carried in the 'pollen baskets' which are located on the bees' hind legs. At this final stage of its life, usually at around six to eight weeks, most worker bees will die in the field – that is, if they haven't already been eaten by a predator or been killed in combat.

The number of worker bees in any one colony will vary throughout the year. During the height of the active season, however, it is estimated that there will be as many as 50,000 to 60,000. The lifespan of the worker bee can be anything from 15 to 38 days, depending on the time of year. In the winter it can survive as long as 140 days as she does not have so much work to do and can live off the stores already built up in the hive.

THE 'WAGGLE DANCE'

Bees are such efficient pollinators because they have learned a sophisticated method of passing information from one to another. As soon as a colony forms, scout bees are out looking for the closest and richest sources of pollen and nectar. When they find a good supply, these scouts return to the nest with samples and they begin to tell the

Below: *The progress of the waggle dance.*

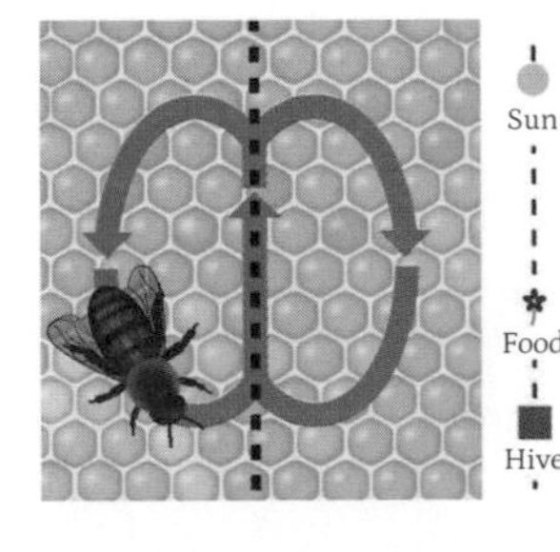

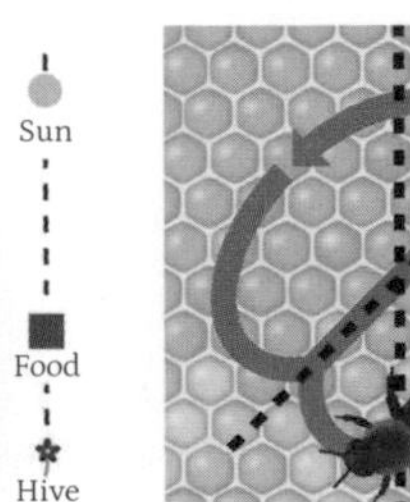

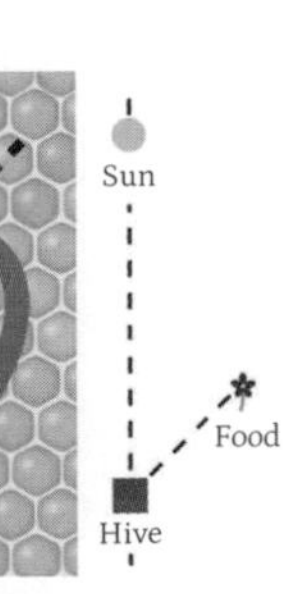

other foragers about the location and how to get there. They do this by performing a symbolic dance language which is based on movement and sound.

The dance is called the 'waggle dance' because the body of the bee waggles from side to side. It takes the form of a figure of eight and is performed by worker bees on the vertical surface of a comb. The worker moves in a straight line in a figure of eight and waggles its body from side to side. When this waggling phase is complete, the bee circles off to one side and returns to the point where it all started. This sequence can be repeated as many as 100 times, with the direction of the final circle alternating each time.

Above: *Drone bees are much bigger and heavier than the queen or worker bees, with a notably thick abdomen.*

It is believed that the number of circles performed is correlated to the size and distance of the food supply. For example, if the worker performs ten cycles in 15 seconds it means that the food is approximately 100 m (330 ft) away. The bee measures the distance in terms of how much energy she has used in travelling. Therefore, the further the foraging site, the longer the duration of the waggle and the bigger the supply the greater number of dance circles.

The dance also transmits details on the direction as well as the distance by using the Sun and gravity. By dancing on top of the vertical honeycomb, she can describe the angle needed to locate the food. An upward tail waggle means 'go towards the Sun', while a downward run means 'fly away from the Sun'. As the position of the Sun changes position throughout the day, so the bees change the angle of their dance, too. While carrying out this intricate dance movement, the bee will often stop and give out small samples of the nectar she has collected to those who are watching. The attending bees are able to glean a good deal of information from this sample, which can also give them a clue to the scent they are looking for.

There is another type of dance that the bee performs and this is one with a circular movement. It is believed that this is to tell the attending bees that there is nectar close to the hive and to go and forage in the surrounding area.

THE DRONES

The drone is the big, butch one in the insect colony and many novices mistake him for the queen. If you look closely you will see that the drone is characterized by the compound eyes that are twice the size of those of the worker bee. Their wings are also longer than those of the queen and hang below the abdomen, which is blunt as opposed to the queen's which is pointed. The drone has no sting and can therefore be handled quite safely.

Drones are the only male bees in the colony and their sole function is to reproduce. Those that do succeed in mating with a queen during her nuptial flight perish during the act.

Their sex is determined by the queen, and should she come across a larger drone cell, she will not fertilize the egg as it passes out, resulting in the drone – this is the result of parthenogenesis.

There are usually several thousand drones in a colony, but their life expectancy is only around 90 days and they can be prone to ejection at any time. The young drones are fed by the workers, but as they reach maturity they feed themselves honey directly from the food stores. Should the food supply diminish for any reason, worker bees waste no time in ejecting the drones from the hive. After the first heavy frosts in autumn, the supply of nectar and pollen decreases, and the colony starts to prepare for winter. They will start to drive the drones out of the hive to give the rest of the colony a better chance of survival.

DRONES AND MATING

Because the drone is built for mating, which is done on the wing, it is essential that he can fly extremely fast. For this reason his flight muscles and wings are much larger than those of the workers. The drone also requires a keen sense of smell to pick up the queen's pheromones and extremely sharp vision to pick up the other drones over large distances. Compared to the queen bee who has 3,000 to 4,000 eye facets and the worker who has 6,900, the drone has as many as 8,600. The drone also has an amazing 30,000 antennal plate organs (sensory organs), compared to the queen with some 1,600 and the worker around 3,000. The drone is fitted with a special odour receptor which enables him to find a queen in flight, and this receptor can track her down from up to 60 metres away. Although the drone is a defenceless bee and is unable to defend the hive, it does have one other purpose to the colony. All bees, regardless of caste, will react when they sense a change in temperature in the brood chamber. Just like the workers they can either huddle together for warmth, or fan their wings to cause a through draught.

Bee anatomy

The bee, like other insects, has three main body regions – the head, the thorax and the abdomen.

THE HEAD

The bee's head is dominated by two large compound eyes. These eyes are capable of seeing ultraviolet light, which is invisible to humans. Between these eyes are three smaller ocelli or simple eyes, which are responsible for registering light levels. Some flowers that appear totally yellow to us, only appear as yellow around the source of nectar to a bee. This has the effect of drawing the bee directly to the important part of the flower. A bee's vision is believed to be sharp for a distance of only about 1 metre.

The antennae are located more or less central to the bee's face. The antennae are controlled by four muscles and it acts as a specialized organ of sense. Much of the communication between bees is done by touching of antennae.

The mouth parts of a bee are far more complex. The mandibles, or jaws, are suspended from the head at the sides of the mouth. These are used to handle objects, manipulate the collecting of pollen and in times of combat with other bees. Positioned just above the mandibles is the mandibular gland which secretes a substance which was once believed to be softened wax. In fact, it is this gland on the queen bee which secretes 'queen substance', the pheromone especially important in the maintenance of colony structure. These glands are almost completely reduced in the drone.

The front part of the mouth is composed of a wide plate called the labrum. It is here that a proboscis (the tube used for feeding on nectar) is formed by bringing together several parts of the lower mouth. The two maxillae and the median labium (a movable flap) form the proboscis, which, when not in use, is folded up underneath the head.

The bee's tongue is covered with fine hairs and the tip is a small spoon-shaped lobe or flagellum that is smooth on the underside, but covered with branched spines on the edges and top. Muscles associated with the tongue allow the bee to 'lap' at liquids and a sucking pump assists when they are feeding. They have special salivary glands which help moisten the food and a special opening in the mouth area for the brood food glands.

THE THORAX

The thorax is the central part of the bee's body, where the legs and wings are attached. Each pair of legs has a different function. The front legs are used to clean the head, eye, mouth and antennae. The middle pair are used to clean the body, secreted from glands in the abdomen. The hind legs are specialized for collecting pollen. Each leg contains long fringed hairs that form the pollen basket. Pollen grain which sticks to the hairs of the body are then brushed back to the inside of the hind leg where they are stored in the pollen basket for taking back to the hive.

There are two sets of wings attached to the thorax. They are membranous and strengthened by veins. The wings are powered by large flight muscles located in the thorax, and when not in flight the wings fold back neatly along the body.

THE ABDOMEN

The abdomen contains the digestive and reproductive organs. It also contains the wax-producing glands which are most productive during the 12th to 18th days of the bee's life. These produce tiny wax scales which are paramount to comb building. On the upper part of the abdomen are seven scent-producing glands. These glands are responsible for producing pheromones which are used at the entrance of the hive or when the bees are swarming, to guide other workers to the right spot.

At the end of the abdomen is the sting which, when not in use, is completely retracted into the body. The sting consists of an upper stylet and two lower lancets. The stylet has a wide bulbous end which connects with the poison sac. When the bee stings, the entire sting apparatus works its way into the wound and continually discharges venom.

SWARMING

Watching honey bees pour out of a hive by their thousands and then forming a great swirling tornado is perhaps one of nature's most spectacular sights. However, a swarm of bees is quite capable of causing the bravest person to suffer a panic attack. Although bees are relatively harmless while swarming, it can still cause alarm to someone who is not used to it and the noise itself can be pretty scary.

So what is swarming all about? It is important to understand that this is a completely natural process and that the bee is only trying to propagate its own species. Swarming generally occurs because of one of the following:

- The colony is opulent – i.e. rich in bees, food stores and general health
- The hive is too cramped and the colony has no room to expand
- The queen is getting old
- The queen is sick or dead
- The hive is diseased and not a safe place for the colony
- Insufficient food supplies

A normal honey beehive will survive the winter with a population of approximately 12,000 bees. The queen will start laying her eggs in January with the purpose of building up a workforce of about 50,000 bees, to ensure maximum productivity during the summer months ready for the forthcoming winter. When a beehive, whether it is in the wild or one belonging to a beekeeper, is running out of room to store honey, the bees know it is time to leave and seek a new nesting site. Experienced beekeepers will be able to anticipate this and supply another box to increase the amount of food storage.

Preparations for swarming usually start about one week before the actual event. The queen will have started to lay fertilized eggs in the queen cells and the workers feed them. These cells are then capped and once they have been sealed off the old queen is ready to leave the hive with approximately half of the workers to set up a colony elsewhere. During this time scout bees will have been inspecting other potential nesting sites.

Once the swarm leaves the hive, it will attach itself to a nearby bush, tree, old chimney etc., and the scout bees will then inspect the sites they have found and return to the swarm. These scouts will dance on the surface of the swarm to give information about the new sites. Eventually the swarm will come to a consensus indicating only one distance and one direction. Because so many workers leave with the swarm, this leaves the remaining colony much smaller which means they will hardly produce any honey the following season.

The swarming season is usually between the end of April and the end of June. Bees will only swarm if there is a good chance that the virgin queens left behind in the hive will be mated, to make sure there are plenty of drones in the area.

Left: *Bees swarm for a number of reasons. This substantial swarm has gathered on the lower limbs of a tree.*

CHAPTER TWO

GETTING STARTED WITH BEEKEEPING

This chapter covers all the essential equipment you will need to get started on this totally absorbing hobby. It is not difficult and the list of requirements is quite small compared to many other pastimes.

The hive

The hive is first on the list of requirements, as without a suitable nesting site you cannot keep bees. It is a good idea to start with at least two hives. If one colony starts to fail, you can use the bees, larvae and eggs from the second colony to help out. If you have only one colony, you will have no backup and it will quickly die out, leaving you with nothing. A modern beehive can look complicated to a novice, but it is just a simple set of boxes stacked one on top of the other. Whether you choose the basic square format or the classically shaped hive, the principle is the same. Let's start from the bottom up and explain each piece in turn.

THE STRUCTURE OF A TYPICAL BEEHIVE

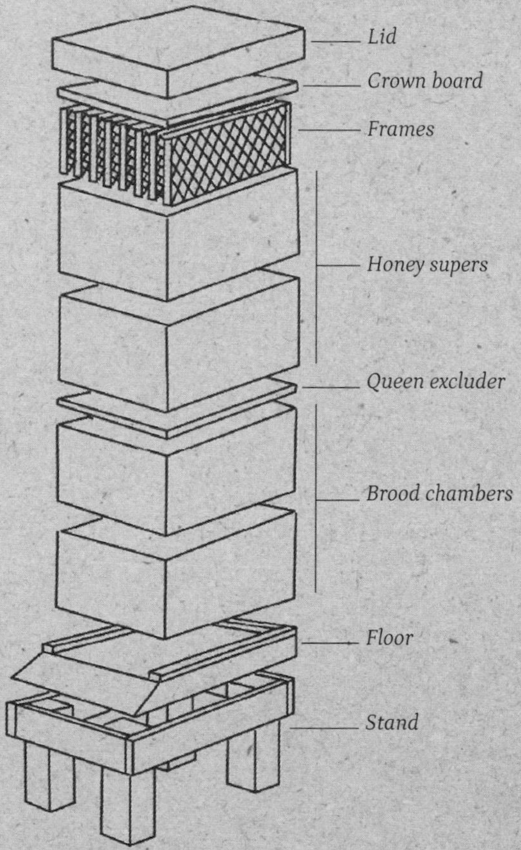

Lid
Crown board
Frames
Honey supers
Queen excluder
Brood chambers
Floor
Stand

Above: *Most modern beehives are those based on variations of a design patented by Rev. Lorenzo Langstroth in the USA in 1852. The most commonly used beehive in the UK is the National but all modern 'framed' hives use the same structure.*

THE STAND

Never stand your hive directly on the ground, as it is essential that damp and weeds do not penetrate the structure. You can either buy a purpose-built stand or you could raise your hive off the ground using pallets, concrete blocks or bricks.

THE FLOOR

The floor is a simple piece of stainless-steel mesh contained within a wooden frame which has three raised sides. The mesh is useful for the hive's ventilation, especially in hot weather. The mesh also helps in the control of the varroa mite, a pest which can be fatal if it takes control in your hive. If the mite drops off the bee it is small enough to drop through the mesh, but will be unable to get back into the hive. Finally, as it is damp and not cold which kills bees, any water that falls on the mesh will drain away immediately.

BROOD BOX (OR CHAMBER)

When you first start out you will probably only use one brood box. This is the chamber where the queen will lay her eggs and where the rearing of the larvae will take place. If the queen in your particular colony is a prolific layer, then you may wish to add another brood box to prevent swarming. By moving the queen excluder up, it will give the queen another area to lay her eggs and prevents overcrowding.

QUEEN EXCLUDER

The queen excluder is a simple grid of slotted plastic or wire that lies on top of the brood box. The plastic ones are preferable as they are not so prone to warping and bending. The slots in the grid are large enough to allow the workers to pass through, but will stop the queen from passing into the higher chambers where the honey is stored. This is important when you come to harvest your honey, as it means you will only extract pure honey when you place your combs in the extractor. It also means that any bees that you accidentally place back into the honey room will not be your precious queen.

HONEY SUPERS

These boxes are super-imposed on top of the brood boxes. You can get both full-sized and three-quarter-sized supers, but it is a good idea to start out with the smaller ones as they are easier and lighter to handle when the combs are full. If you find your bees are filling your supers quite quickly, you can always add another one to your stack.

Below: *A wooden frame bearing a sheet of fresh beeswax that the bees will add to as they start to make their own wax.*

FRAMES

The frames are one of the most important parts of the hive. They can be made of either wood or plastic and they are the frames which hold the beeswax. You can buy wooden frames with beeswax already stamped into hexagonal shapes. These wax sheets are held in place by thin wires

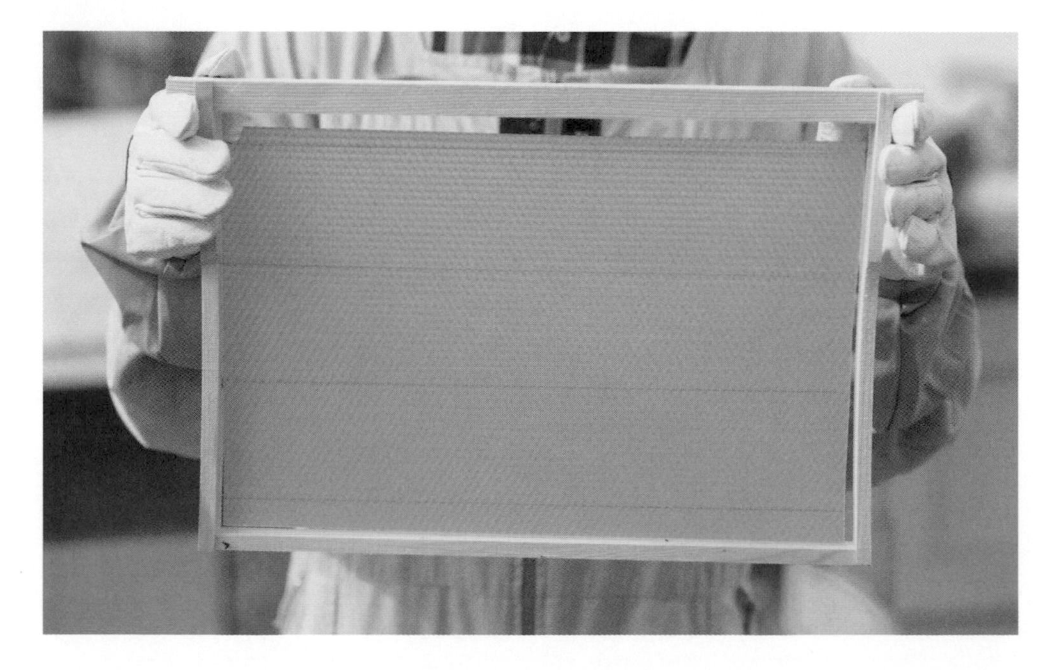

which cross the frame. This sheet of wax acts as a starter for the bees to make their own wax, which they then form into perfect hexagonal chambers. Again, for the beginner, plastic frames are ideal as they are both lighter and easier to clean. They are also relatively strong which means there is less risk of damage when extracting honey. The plastic is already formed into the hexagonal cells and many beekeepers simply dip the frames into molten wax to give their bees a start. Whether you choose the plastic or wooden versions, frames come in a variety of shapes and sizes. You will need to make sure that the frames you buy will fit your choice of hive.

SPACERS

When you hang the frames within the box you must make sure that there is enough 'bee space' between each one. This space is simply the amount of room a bee needs to pass between the frames easily. The recommended space required is approximately 7.5 mm. If the space is not large enough the bees will fill the gap with propolis (bee glue), and if it is too large they will fill it with brace comb.

Spacing can be built in as part of the frame in some of the well-known brands, or separate spacers can be nailed into the brood box or super, and the frames simply fit into them. You can also buy metal or plastic ends that fit onto the ends of the frames themselves. This spacing applies to any part of the hive where you have frames.

FEEDERS

Feeders are plastic frames with sides and an open top. They are used to provide your bees with sugar syrup, when required. All you need to do is remove one outer frame in the brood box and replace it with a feeder. Before you fill the feeder with the sugar syrup, it is important to add some material such as bits of wood or dried bracken, so that the bees will have something to hold on to while they are lapping the solution, otherwise they could drown in the sweet liquid. There are other types of feeders, but frame feeders are by far the best for the beginner.

Another form of feeder is the bucket feeder, which consists of a small plastic bucket with a tight-fitting lid that has been punctured with lots of tiny holes. The bucket is filled with sugar syrup and inverted over the frames. You will need a spare box to surround the feeder and also a hive lid to put on this box. These bucket feeders are ideal if you want to give your bees a slow feed. They are very easy to use and cheap to buy, but are definitely not as convenient as the frame feeder.

Many commercial beekeepers invest in a lid feeder that not only holds more syrup but can be easily accessed without having to open the hive.

Right: *Sheets of foundation wax stamped with hexagonal shapes.*

FOUNDATION WAX

If you are not purchasing a hive containing bees, you must provide your new colony with some foundation wax. This wax will be used as a basis for the honeycomb and brood chamber. Wax sheets are thin and can easily get damaged, especially when you start to harvest your honey, so you need to handle them with care. Foundation wax is important because:

- It encourages the bees to build straight combs within the frames.
- It saves on the bees' labour time in the construction of their combs.
- It helps when you come to extract the honey as the frames are reinforced with wire.

THE CROWN BOARD

The crown board or inner cover is not an essential part of the hive, but some beekeepers find them useful when they use a bucket feeder as they normally have a round hole in them. The crown board sits on top of your hive immediately under the roof. In the summer it is likely to be on top of a super, but during the winter months it will sit on top of your brood box.

THE LID

Lids come in a variety of designs but they all serve the same purpose. You can have a gabled roof, or a simple flat roof that fits right over the hive at the front and back. The easiest types are the flat telescopic lids that stay on in strong winds and allow you to stack the hives on top of each other for ease of moving.

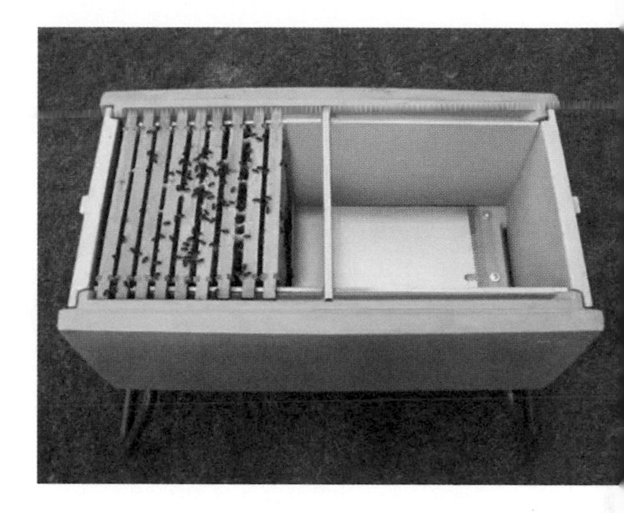

Above: *A Beehaus seen from above, showing how the frames are stacked horizontally, rather than vertically, as in a conventional hive.*

Some of the flat lids come with a cavity that can be filled with sugar syrup for feeding. This is an advantage because you can access it without opening the hive which is less intrusive to the bees. Whatever type of lid you choose, it is essential that it is watertight but still allows enough ventilation to get into the hive. If you live in a hot area, then it is advisable to paint the lid white, and the rest of the hive for that matter, to try to reflect some of the heat.

THE BEEHAUS

To attempt to make beekeeping more accessible to city dwellers, a new plastic beehive has been introduced on the market called the 'Beehaus'. It resembles a giant coolbox which stands at waist height, with a lid that can easily be removed for inspection and maintenance of the hive. It is almost twice the size of a conventional hive, giving the bees plenty of room for expansion and lessening the chances of swarming. It is also easy to clean and impervious to woodpeckers, which do have a habit of penetrating wooden hives. It comprises three layers of plastic that are separated by air pockets, allowing the bees to regulate the internal temperature more easily. The other advantage is that it is light, easy to move and ideal for those people who want to keep bees on a shed or garage roof.

BUYING SECOND-HAND HIVES

An excellent way of starting out on your new hobby is by buying second-hand equipment. This can be obtained from a local beekeeping association or keep your eye open for local auctions. Second-hand equipment is usually much cheaper than buying new and, as long as the hive is in good condition with no rotting wood, it is the perfect way to get started. One disadvantage of buying a second-hand hive is that it could harbour disease.

The safest way of making sure there are no spores in your new hive is to run a flame gun over the surfaces, making sure you go into all the cracks and corners. Alternatively, you can soak the woodwork in potassium hypochlorite. You will need to wear protective clothing to do this and also have a very large tank to immerse the hive. Scorching is definitely the easier method, but make sure you don't set fire to the hive.

Other tools and equipment

As you become more involved with your new hobby you will probably want to try some of the new and more interesting tools available for beekeeping, but to start off with you only really need two – a hive tool and a smoker. This section gives you details about all the types of equipment you can buy so it is up to you just how much money you would like to spend.

THE HIVE TOOL

These handy little gadgets are a must for any beekeeper. Hive tools are specially designed levers with scraping edges and other features that aid in the removal of frames from beehives. They can also be used to scrape the build-up of propolis or brace comb from hive parts.

They come in three basic designs. There is a small, narrow hive tool and frame lifter, which is handy as it can be held in the little finger of one hand, leaving you free to lift the frames. The curved portion gives rise to the name 'J' hive tool, and is used as a hook to lever out a frame using the shoulder to rest on an adjacent frame.

There is another 'J'-shaped tool, but it is broader and stronger. The curved portion is used in the same way as the first example. This type is ideal for inserting between frames and, using a twisting motion, it can separate and break any propolis that is sticking the frames together. The third type, pictured here, is broad at both ends with one curled round like a scroll.

Below: *Various useful beekeeping tools including a hive tool, fork and scraper.*

To separate the frames, you drop the curved portion of the hive tool between adjacent frame top bars and rotate the tool to left or right. A levering action can also be performed by lipping the curved portion under a lug and rocking the tool backwards so that the rounded portion rests on the adjacent lug. The unusual-shaped hole in the top is intended for removing nails, should you need to do any quick hive repairs during an inspection. Alternatively, you could use the hole to attach a piece of string so that you can hang your hive tool in a convenient place. It is advisable to have more than one of these inexpensive tools as you will find you probably lose one on a regular basis, especially when you spot the queen and just drop the hive tool in the grass.

Above: *The smoker is used to control bees in a number of circumstances including swarming.*

THE SMOKER

A bee smoker is a tool used to blow smoke into a beehive before inspecting, manipulating or handling the hive. Once calmed by the smoke the bees are less likely to sting. The original smoker, which was invented by Moses Quinby of New York in 1875, consisted of a firepot, bellows and a nozzle to direct the smoke. The bellows force air through the firepot, pushing the smoke out of the nozzle.

Although the secret of smoking bees has been known for thousands of years, the actual scientific reason as to why it calms the bees has only been explained in more recent years. Under normal circumstances, if a beehive is threatened, the guard bees will release a volatile pheromone substance called isopentyl acetate, which is better known as 'alarm odour'. This will alert the older bees within the hives – the ones with more venom – to defend the hive by attacking the intruder. If smoke is blown into the hive first, the guard bees' receptors are dulled and they fail to sound their alarm. The smoke also has a secondary effect in that it causes the other bees to instinctively gorge themselves on honey.

Fearing that the hive is about to be burned down, they fill their stomachs before getting ready to evacuate. This gorging has a tendency to pacify the bees. When you buy a smoker, choose one with a protective grid around the outside so that if you pick it up without gloves, you will not burn yourself. It is also worth spending a bit more money on a large smoker, as the small ones do not hold enough fuel and you might find you have run out of smoke before you have finished. The bellows also need to be made of strong material, as this is one of the first parts of the smoker to perish.

Your aim is to produce a cool, dense smoke and the ideal fuel for this is either corrugated paper, old egg boxes (not the plastic type), old sacking or dried grass. Make sure your smoker is burning well before you approach your hives – you don't want to be stuck with a lot of angry bees and a smoker with no smoke. Many experienced beekeepers find they can predict the mood of their bees and say they do not need to use smokers every time they handle their hives.

FRAME REST

If your hive does not come fitted with a frame rest, then this is probably another essential piece of equipment. It is literally somewhere to hang your first frame to give you room to manoeuvre the others.

The next few items are by no means essentials, but they could be handy and are worth investing in if money is no object.

Bee brush

This is exactly what it says, a soft brush with a fairly long handle which is perfect for gently brushing bees from a comb.

Wire excluder cleaner

Some beekeepers use a wire brush to clean their excluders, but there are handy little stainless steel combs available, which are a very effective way of cleaning them.

Floor scraper

This is another handy little tool that has a stainless steel shaft and blade with an easygrip rubber handle. It is invaluable for removing floor debris and wax moth from under the mesh floor.

Manipulation cloth

As you will not want to leave all your frames exposed while you are working on your hive, a manipulation cloth is handy for laying on the part not in use. It is heavy enough not to be blown away in the wind, and can be folded back to allow a frame to be taken out.

Frame cleaner

A frame cleaner is a tool with a swan-shaped neck for cleaning frame grooves. It can get into all those difficult places.

Queen catcher
As marking the queen is an important process in hive maintenance, you will need to invest in a queen catcher. Its name says exactly what it is, and one of the best types is like a large, plastic bulldog clip. You open the catcher by pressing the sprung levers between the thumb and forefinger. Once open it can be placed over the top of the queen and closed around her without causing any harm. It doesn't matter if you catch a few workers at the same time, as the slots in the catcher are large enough to allow the workers to escape, but not sufficient to release the queen.

Marking cage
This little cage is ideal for holding and marking the queen. It is a little round cylinder with a mesh end and a plunger which is padded with sponge so as not to damage your precious queen. Once the queen is inside you can use the plunger to encourage her to go to the top of the cage, where it is easy to mark her through the mesh.

Marking fluid or pens
You will learn about marking the queen later in the book as this is a very important part of beekeeping. The beekeeper needs to know the age of the queen in each hive to keep an eye on her laying abilities. The fluid or pens are colour-coded and are quick-drying, water-based and non-toxic.

Below: *One design for a queen catcher which operates a bit like a bulldog clip. Less experienced beekeepers are advised to wear gloves when close to bees.*

Equipment for extracting honey

Before you can start extracting your honey you will need to buy some extra equipment. The main piece of equipment is the extractor itself, but you will also need an uncapping knife or machine, filters and containers for holding the honey.

Below: *Frames in the frame basket of a honey extractor. They will be spun round to pull out the honey from the honeycomb.*

HONEY EXTRACTOR

A honey extractor is a mechanical device used to extract honey from the honeycomb without destroying the comb itself. It is a very uncomplicated piece of machinery which works by centrifugal force. A drum or container holds a frame basket in which the honey frames are placed. You then spin the basket and the honey is literally flung out to be caught in the base of the drum. This method means that the wax comb stays intact within the frame and can be reused by the bees. There

are two types of extractor, the radial or tangential, depending on how the frames are put into the basket. Radial baskets have the top bar of the frame facing outwards, while tangential baskets have one side of the basket facing outwards

It is important when you spin the frames that the honey is extracted from both sides. It is easier to do this with a radial extractor because you simply rotate the basket in the opposite direction. With the tangential extractor, you need to actually turn the frames around. Small extractors can either be turned by hand or run by an electric motor.

Extractors should be made from stainless steel or a high grade plastic, as many of the old tin-plate extractors are no longer considered to be food-grade quality.

During the extraction process, the honey is forced out of the uncapped wax cells. The honey runs down the walls of the extractor and gathers in pools at the bottom. A tap or honey pump at the bottom of the extractor makes collecting the honey an easy task. You can either obtain extractors from bee-equipment shops or second hand from association auctions. It is worth reading through advertisements in beekeeping magazines when you are thinking of buying this type of equipment. Alternatively, if you already know someone who owns an extractor, then it might be a good idea to borrow it until you get used to the process of extracting your own honey.

UNCAPPING KNIFE

The honey bee will cap the cells of the honeycomb with wax once the honey is the right consistency. Before you put your honeycomb into an extractor, you will need to remove these caps. A bread knife with a serrated edge works very well, but many beekeepers find a purpose-made uncapping knife easier to use.

Uncapping knives are best used when they are hot as they will slice through the wax more proficiently. You can either dip your knife into a bowl of hot water or, alternatively, you could buy an electrically heated model to make life easier.

If you are really serious about producing honey on a large scale then you might want to invest in an electric uncapping machine, which has a revolving nylon brush that spins rapidly to remove the capping.

HONEY FILTERS

When you place your honeycomb in the extractor you will find it extracts more than just the honey. The liquid in the bottom of the extractor will contain pieces of old comb, bits of bee, twigs, pieces of broken frame and so on, and you will need to pass the honey through a filter before bottling.

A filter could be a simple muslin bag, a clean pair of old nylon tights or even a fine kitchen sieve.

If you want something a little more sophisticated you could buy a bucket filter. You simply place the bucket filter in a 5-gallon bucket and pour the honey through the filter. The filters come in various meshes, and they are easy to wash and can be used over and over again.

HONEY TANKS

Once you have extracted and filtered the honey you will need some plastic honey tanks for storage. These come with snap-on lids to keep out any debris and many come with valves at the bottom for ease of bottling. By leaving the honey to settle in the tank for a day or so, not only do the air bubbles rise to the surface, but specks of pollen will also settle, leaving your honey clear and ready for bottling.

Above: *Filtering extracted honey through a sieve will remove impurities.*

SUITABLE CLOTHING

The type of clothing that a beekeeper chooses is very important. Whether you are allergic to bee stings or not, it is never a pleasant experience to feel bees crawling inside your clothes, especially around your face. It is advisable to wear light-coloured clothing, as bees tend to be drawn towards darker colours, and it also makes it easier to see whether you have any bees left on you when you come to de-robe. By far the best form of clothing is the all-in-one suit which comes without intrusive pockets, has elasticated wrists and ankles and a veil that is incorporated into the suit. The separate veils that have drawstrings or simply tuck into your top are not advisable as the bees can easily find their way inside these.

A good-quality suit is not only easy to take on and off, but it affords you far more protection. Although the initial cost may be a little high, you can guarantee that it will last you for several years. If you are really concerned about getting stung then you should purchase one with a tougher fabric. They come in several thicknesses, and the thicker ones mean that the bees have less chance of penetrating the material.

The ideal footwear is a pair of wellington or rubber boots which you should wear with your trousers tucked in. Gloves are also an asset and if you don't want to go to the expense of buying purpose-made gloves with gauntlets, then a pair of household rubber gloves will do the trick. Although many beekeepers prefer to work without gloves because they claim they cannot 'feel' what they are doing, this is fine if you really know your swarm but it is not recommended for the beginner. Washing-up gloves will split fairly regularly, so you might like to buy a good pair of leather gloves with gauntlets.

Below: *Especially when starting out it is worth investing in a well-designed beekeeping suit and gloves.*

Getting your bees

Obtaining bees is not difficult and can be accomplished in several ways. The easiest, and sometimes the best, way to start is by purchasing established colonies from a reputable local beekeeper.

The best time of year to obtain your bees is in early spring. At this time of year your colony will be small and gentle and it will give you time to acclimatize yourself to your bees. Also, around June, flowers are in full bloom and your colony should have built up sufficiently to take advantage of this harvest.

If you have already purchased your beehive from a local beekeeper, it may have come with bees already installed. Buying two colonies instead of one is a good idea, as it allows you to interchange frames of brood and honey if one colony becomes weaker than the other and needs a boost. It is essential to carry out some form of check on the colony – has it got a queen, is it disease-free and has it got a sufficient build-up of food for the colony to survive? If you decide to buy your bees from an auction, then there should be an inspector present who will be able to check the colony before you take it away. Although buying an established colony may be easiest, it is worth mentioning that this colony will come with its own guard bees ready to defend the nest, which may be a bit daunting for the newcomer. For that reason you may prefer to buy a nucleus of bees (or nuc).

Above: *Bees, including a queen bee, can be bought and installed as a nucleus.*

BUYING A NUCLEUS

This is the usual method of obtaining bees in the United Kingdom and Europe. A nucleus is a box which contains four or five frames – one or two brood frames, one or two frames of honey and a frame of comb or foundation wax. You will need to let your supplier know the size of your hive before buying these frames, to ensure they fit. The advantage of starting with a nucleus is that the queen will have already laid eggs, and that it will have a brood ready to emerge, helping to build up your colony. Many beginners prefer to start this way, gradually building up to a full colony.

Right: *A package of bees being inspected prior to installation.*

BUYING A PACKAGE

This is the preferred method in the USA. The bees are sold by the pound in weight and come in wire-covered boxes together with a queen. The package usually consists of anything from 8,000 to 12,000 bees and arrives with a frame full of sugar syrup which feeds the bees while they are in transit. The queen is in a separate wooden cage with one screened side through which she feeds, which means she is well protected during the journey. The main disadvantage of this method of buying bees is that it will take approximately three to four weeks before any brood emerge. The bees will need to be put in their new hive on the day of their arrival, preferably in the late afternoon just before dusk.

GENTLE BEES

If you are buying a hive that already has a swarm, then you probably won't get a choice as to the type of bee you have. Certain strains of bees are far more aggressive than others and it goes without saying that a beginner would rather work with gentle bees. The ones that are known for their non-aggressive behaviour are the Italian or Cecropian bees, while the Spanish or Iberian bee is the most savage after the infamous killer bees of Africa. Many breeders will produce gentle bees for you to buy as a nucleus. Your local beekeeping association will recommend suppliers.

CATCHING SWARMS

This is an interesting, and obviously cheap, way to start, but an inexperienced beekeeper is not advised to do this without help, even though bees are at their gentlest when swarming. The easiest way to obtain a swarm is to contact your local beekeeping association and to ask if there is a swarm in your area. If, however, you do have some experience of handling bees and wish to attempt to capture your own swarm, here is a list of essential equipment:

- A container for the swarm – a bee skep is ideal for this. A cardboard box is a cheap alternative. Rubbing the inside of the container with the herb lemon balm is said to make it attractive to bees.
- A piece of net curtain large enough to stand the inverted container on and secure by tying or wrapping over the top.
- A small block of wood to prop up the container.
- Secateurs and pruning saw.
- A length of rope to pull branches down and a weight.
- A ladder is a useful piece of equipment if you have room to carry one.
- Your protective clothing and smoker.
- A queen cage will be useful if you manage to see the queen.
- Telephone numbers of local beekeepers in case you need assistance.

Above: *When dealing with a swarm, it's worth asking a number of questions so that you are able to deal with it in the most appropriate way.*

When you receive the initial phone call regarding the swarm try to gather as much information as possible.

- Where are they?
- How high up?
- How long have they been there?
- Are they on your property?
- Ask for a description – in case they are bumblebees.

First, remember your own safety and do not take risks. A skep full of bees can weigh several kilograms and could easily knock you off balance if you are up a ladder. Never attempt to remove bees from places that are too high to reach, for example from chimneys or very high branches.

There are four main methods of capturing swarms and, whichever method you use, make sure than any spectators stand well back. They are obviously going to be interested in what you are doing because a swarm of bees in formation makes quite a spectacular, if scary, sight.

Method 1: Shake the swarm into the container

If you are really lucky, the bees will have clustered on a single branch that is close to the ground. Spread your sheet on the ground and have your smoker alight and to hand. Holding the container directly under the swarm, give the branch a sharp jerk. This should dislodge most of the bees and hopefully this will include the queen. Slowly invert the container in the middle of the sheet and prop up one side. In a very short space of time, you will see the bees at the opening performing their duty of attracting the other bees to join them by fanning their wings. You can use a little smoke to push any stragglers inside. Smoke the branch the swarm was on to mask the scent of the queen, otherwise this will continue to attract other bees.

If you were unfortunate enough to have missed the queen, the bees will not stay and will return to the branch. It is advisable to leave the swarm until early evening when hopefully all the foragers will have returned. If the branch is too high to reach comfortably, then a weighted rope thrown over the branch can often lower it enough to enable you to reach it without the assistance of a ladder.

Method 2: Smoking the swarm into the container

If the swarm is in a position where you cannot shake them into the container, you can persuade them into a container placed above them by using smoke. Bees always tend to walk upwards into a darkened space. Don't use too much smoke;, a little at the base of the swarm is usually sufficient to make them move. Make sure the box is firmly secured and will not fall when it is full of bees. Once you have coaxed all the bees into the container, invert it as before. This process could take a while, so there is no point in being in a hurry.

Method 3: Brushing the swarm into the container

A swarm that has decided to settle on a wall can be brushed into a container. This is when your cardboard box comes in handy, as the long flat side can be laid against the wall. Using your gloved hands, a large feather or bee brush, gently brush the bees into the container and secure firmly.

Method 4: Using a frame of comb

If all the above methods have failed, then you can try placing a frame of comb close to the bees. The bees should soon cover the frame and these can be shaken into the container and the frame replaced. If you have a brood frame, this will increase your chances of attracting the swarm. As with all these methods, once you have attracted the queen, the remainder of the swarm will follow.

Hiving a swarm

Hiving a swarm is not unlike hiving a package of bees. If the swarm is close enough to the ground, you could take a super that contains old combs underneath the swarm. The odour from the bees that previously occupied the hive will attract the scouts, as scent plays an important role in the social organization of the colony. Bend the branch containing the swarm over the top of the super and dislodge the bees with one or two hard shakes. If the swarm is higher up the tree, you can either use a ladder and carry the super to the swarm or you can cut the branch with the bees intact as a group. Then you simply shake the branch into the open super without the risk of falling.

Occasionally the swarm will not accept their new home. They will leave the hive en masse and regroup in a new swarm somewhere else. Very often, if the bees have stayed in the vicinity, the second attempt at rehousing them is more successful. Waiting until late afternoon is a good idea, as the bees are less likely to wander after it gets dark.

Below: *The old-fashioned straw skep, still used as a hive, is also very useful for hiving a swarm.*

Positioning your hives

The most frequent question asked by people starting out is – Where is the best place for my hive?

A good site is somewhere that is secluded, exposed to full sunlight (dappled shade is fine), has good air circulation and water drainage, a source of fresh water, out of the wind and is in close proximity to plenty of flowering plants. It is also helpful if there is a small shed or building nearby where the beekeeper can keep his equipment. Try not to place a hive under a tree where it can be dripped on during and after rain. Also try not to place hives near any overhead power cables and keep the immediate area clear of tall grass and weeds.

It is a good idea to try to keep the hive out of the public eye as some people are afraid of bees and others may be tempted to vandalize or even steal your hives. You need to leave plenty of room around the hive so that you can work comfortably without knocking into things. It is also quite a good idea to place a hive next to a high hedge or fence, which encourages the bees to fly up and over the hedge to forage.

MINIMIZING AND PREVENTING DRIFT

Drifting is a problem that is associated with many apiary sites. Drifting occurs when a bee mistakes another hive for its home. If the rogue bee is carrying nectar, pollen or water, the guard bees at the entrance to the hive will normally allow the stranger to enter. It has been discovered that hives at the end of a row will often have larger populations and gather more honey than other hives in the same row. The reason – they pick up drifting bees from other hives. It is essential to try to minimize drifting because it means that some of your hives could have a reduced number of foragers. This can affect their ability to gather honey and can also lead to the spread of disease. If you are only starting out with a couple of hives, drifting will not be a problem, but if you decide to expand, this could become a major consideration.

You can do several things to try and eradicate the problem of drifting, something which is overlooked by many beekeepers.

- Arrange your hives in an irregular way. By placing your hives in amongst trees or shrubs, with the entrances facing in different directions, the bees will be able to use the foliage as landmarks and correctly identify their own hive.
- If you do not have any natural landmarks, place or grow some to assist the bees in their navigation.
- Arrange your hives in wavy lines or a horseshoe pattern, not straight lines.

Right: *These hives are part of a larger apiary in a field. The siting will have been subject to a number of considerations including aspect, the type of foraging plants available and privacy.*

Above: *The centre of a sunflower is packed with pollen-bearing stamens and this worker already carries a heavy load of pollen on her legs.*

- Arrange your hives in pairs at least two to three metres apart.
- Paint different coloured shapes on the entrance to the hives.
- Make sure there is at least three metres between rows of hives.
- Avoid placing the hives where they might be affected by wind.

THE IMPORTANCE OF POLLEN

Without an abundance of pollen- and nectar-producing plants, beekeeping is not possible anywhere. A colony of bees needs to collect a surplus of honey to ensure its survival through the winter, and pollen is its source of protein. Approximately one cell of honey and one cell of nectar is needed to produce just one new bee. Although bees can fly as far afield as 12 km (7.5 miles) to collect pollen and nectar, research has shown that it is those foragers who gather most of their food within a 1-km radius that prosper the most. Compared with other insects, bees are extremely hairy. Each hair is designed so that it is highly effective at catching pollen. To make sure the pollen doesn't fall off, the bee will regurgitate some nectar and mix it with the pollen. When the bee lands on the next plant, some of the pollen left on the bee's body hairs will fall off, enough to pollinate the new flower.

THE IMPORTANCE OF WATER

All bees, regardless of where they are kept, require a constant source of water. If there is no water supply close to their hive they will seek out a nearby river or pond, or even the neighbour's swimming pool, so it is a good idea to make sure they have an adequate supply of fresh water. Water is essential to the bee in maintaining temperature and controlling humidity in the hive, and brood-rearing and the dilution of food sources.

The task of some honey bees is to carry water back to the hive. One bee may make as many as 50 trips each day, each time collecting around 25 mg of water. When the colony is short of water, the foragers will divert from collecting nectar and pollen to join in the effort.

In a very hot climate, one colony of bees will need several litres of water every day just to survive. It is, therefore, the beekeeper's responsibility to provide this water for his colony. The container should be at least one metre in diameter, as the bee finds the water supply by the increased humidity in the air above the water. It is a good idea to provide floats – i.e. pieces of cork – to prevent the bees from becoming waterlogged while lapping.

Below: *Even a drop of dew can provide crucial sustenance to a thirsty honey bee.*

Urban bees

Perhaps it might seem strange to think of people keeping bees in the middle of a city, but in fact some of the best honey comes from hives in an urban park or garden. Surprisingly, these areas are richer in nectar and pollen than the countryside. City dwellers must take special care so their bees do not become a nuisance to neighbours, or even appear to be a problem at all. Bee stings are usually the biggest concern of those living next door to a beekeeper, so it is important to minimize the risk. Perhaps you could keep them sweet by giving them the odd jar of honey at the end of the season.

There are two ideal sites for urban bees – a rooftop terrace or an enclosed garden.

THE ROOFTOP TERRACE

Innovative new approaches are being tried to reverse the decline in the number of honey bees. Among them is the siting of beehives on rooftops. These are an ideal situation, offering a high flightpath and a home that is away from the public's gaze. Urban bees and home-grown honey are definitely a rising trend and even some big-city hotels and large stores are participating in this conservation strategy.

ENCLOSED GARDENS

If you are thinking of keeping bees in an urban garden, the area should be as enclosed as possible. High walls and fences should be in front of and behind the hives, to encourage the bees to fly high enough so as not to annoy the neighbours.

So whether you choose to keep your bees on your roof terrace, your balcony or in a small, enclosed garden, there is no reason why the city dweller shouldn't enjoy the benefits. After all, beekeeping doesn't need to be the reserve of country folk and we have already shown that you don't need a lot of space to keep them. Because of the wide diversity of plants and flowers grown in the cities, the bees love the variety and the result is nearly always gorgeous-tasting honey. Just make sure that, like the country bees, townies have plenty of water.

AVOIDING COMPLAINTS

To make sure that you minimize any complaints from neighbours, here are a few simple rules to follow:

- Try not to be too ambitious – limit your hives to just two or three at the most.

Right: *Two small hives on a balcony allow a city beekeeper to pursue their hobby even in town.*

FLOW

- Provide a constant supply of water so that the bees do not wander to your neighbour's swimming pool or pond.
- When acquiring bees go to a reputable dealer who is known for supplying non-aggressive bees.
- Do everything you can to minimize the chances of swarming.
- Try to place your hive in a sheltered position out of the view of your neighbours.
- Erect a high fence or hedge to force the bees' flight path above people's heads. Bees will normally travel in a straight path to their hive, and a fence raises their flight path and reduces the chance of bees accidentally colliding with someone walking nearby.
- Speak openly about the advantages of bees and how essential they are for pollination even in city areas.
- Make sure you let your neighbours have some samples of your produce.

The bee's defensive reaction is greatly influenced by environmental conditions. It is a good idea if you do not place your hive in a direct line with your neighbour's porch or security light. Although bees do not generally fly at night, if they are disturbed by something they could be attracted by the light and cause a problem next door. It is also advisable not to work on your hive if your neighbours are having a barbecue or mowing their lawn, as the extra activity could make them angry. Your main aim if you want to keep bees in an urban area is to try to make your friends and neighbours feel safe and ensure your hobby will not be a nuisance to them. Take time explaining how the hive works and hopefully you will not get too much opposition.

INSURANCE

It is essential to have insurance if you are thinking of keeping bees. The easiest way round this is to join a national or regional beekeepers' association whose members are covered by a public liability insurance which provides protection against claims arising out of accidents caused by the member's beekeeping activities.

The same organization should also be able to help you arrange insurance against malicious damage and theft of beehives. Although bees kept in the countryside are not likely to cause offence to anyone, they could be stolen or vandalized so insurance is very important.

The day your bees arrive

It is early spring, you have acquired your equipment, positioned it in the best place possible, bought the appropriate clothing and by now you will be keen to actually get started. One final job before collecting your bees is to make up some sugar syrup – you will need 2 litres per hive. This can be made by mixing 1 kg of white sugar with 1 litre of water and should be stirred well until it is clear. For the fastest development of a colony, it is best to use invert sugar and a recipe for this can be found on page 104, but regular sugar syrup will suffice for the initial feed.

THE HIVE AND BEES

If you have decided to go for the whole package – a hive full of bees – then your supplier will hopefully have already done a check with you that the queen is present and laying and that there is enough food to last them a few days. It is best to start with Italian bees as these are gentle, good foragers and disease resistant. When the hives arrive – an exciting moment for the first-time beekeeper – place them in your prepared site. It is recommended that you face the entrance south if you live in the Northern hemisphere and north-east if you live in the Southern hemisphere. Although this is by no means essential, it is worth bearing in mind that the early morning sun warming the entrance to the hive will encourage your foragers to make an early start.

Make sure the hive stands off the ground by either placing it on its stand, a pallet, some bricks or anything that makes a firm foundation. Use a small wedge at the back of the hive to slightly tilt it forwards. This prevents any rain from getting in – damp is fatal to bees. Then unblock the entrance and leave the colony for a few days to settle down before carrying out any further inspections.

INSTALLING PACKAGED BEES

If you have purchased a package of bees, the procedure is slightly more complicated but still fascinating for the novice. When your package arrives you will need to place it in a cool, dark room. The ideal temperature should be between 18 and 20°C. Using a brush, sprinkle some sugar syrup over the surface of the screen. This helps to calm the bees and makes them easier to handle. Continue this procedure until the bees stop feeding.

Put on your protective clothing and place a bucket of sugar syrup close to the entrance to the hive. The best time for placing your bees is early evening or late afternoon when the bees are in a more dormant state. Other

bees are less inclined to rob the colony at this time of day. The other reason for choosing this time of day is because the bees may become disorientated by their new location and could either become lost or drift to another hive. If you are installing more than one package make sure your hives are several metres apart and it is quite a good idea to paint the hives different colours to prevent drifting (see page 50).

It is advisable to install packaged bees on frames with new foundation, not old comb. The reason for this is because if the bees are carrying any honey with disease spores, the honey will be used in the construction of new wax and the disease spores will be lost. If you install your bees on to old combs, there is always the chance that the bees will deposit any honey they carry into the cells. This could later infect the entire colony. Reduce the entrance to the hive by using a purchased entrance reducer or a small bar of wood, so that only one bee can get in and out at a time. Remove half of the frames from your brood chamber.

Take your package to the hive and then rap it on the ground to knock the bees to the bottom of the box. Carefully remove the cage's cover and remove the feeder which was supplied with the bees. Again, knock the bees to the bottom and then remove the queen's cage and replace the lid. Check that the queen is alive and, using a nail, puncture the candy plug in the queen's cage to ensure the worker bees can release their queen by chewing through. Next, wedge the queen's cage between two frames within the hive, candy end up. Return to the box of bees and once again knock the bees to the bottom and remove the cover. Pour half the bees over the queen's cage and the remainder into empty spaces within the chamber. Leave the package on the ground in front of the hive's entrance, so that any remaining bees can find the hive. As the bees slowly spread throughout the hive, gently replace the frames you removed earlier.

Being careful not to crush any of the bees, replace the inner and outer covers back on the hive. Make sure you have filled the feeder frame so that the new colony has plenty of food. If you do have any problems getting the bees down inside the frames, a small puff of smoke will soon send them scurrying for safety.

Wait for a few days and then check to see if the bees have released the queen from her cage. If she has been released, then you will probably find her walking slowly on one of the centre combs. If the bees have failed to release her, return the cage to the hive until she is released. A week after the release of the queen, check the colony again. By this time you should find white wax combs under construction with cells containing syrup, eggs or even young larvae. If you do not find any eggs it might mean that the queen is dead and she will need to be replaced immediately.

Left: *Installing a new package of bees into the hive is a delicate operation.*

INSTALLING FROM A NUCLEUS

You will probably find every beekeeper has a slightly different method of installing bees from a nucleus, but if you follow these basic instructions you won't go far wrong. The main thing to remember is to try not to let the bees get cold as you move them from the nuc box into their new home. It is best if you wait for a warm day when there isn't too much wind and make sure all your equipment is assembled or to hand before you open the nuc. This is also the time to light your smoker.

Place your new super (brood box) on its stand and bottom board, but do not place any frames in yet. Place the nuc box right beside the hive in order to lessen the chance of losing your queen while you are moving the frames.

Carefully lift the lid off the nuc box and give a little puff of smoke across the top of the frames, to help calm the bees. Using your hive tool, gently loosen the frames so that they can be lifted out easily. It is easiest to start with the second frame rather than the outside one which is up against the wall of the box. Lift the frames out of the nuc box one by one, carefully transferring them to the empty super. As you do this keep a watchful eye out for the queen, but don't take too long as you risk chilling the brood. It is important to remember to place the frames in the super in the same order they were in the nuc box. Some beekeepers prefer to place the frames in the centre of the brood box, while others like to put them to one side. If you do start from one side, they are only exposed to cold air on one side, the other being insulated by the wood of the box. Fill the remainder of the brood box with frames. If you do have any frames of drawn comb that are disease-free and in good condition, it is a good idea to use them.

That will save the bees a lot of hard work and the colony will build up more quickly. If you do not have any, and you are only just starting out, it is fine to fill the brood box with frames of new foundation. You will find that there will be a lot of bees still in the nuc box. Turn the box upside down over the new hive and give it a few sharp shakes. This will cause most of the remaining bees to drop into the hive. Any stubborn bees left in the box will soon find their way to the queen if you leave the open nuc box just in front of the hive entrance.

Reduce the size of the entrance so the bees have an easier time of keeping themselves warm. If it is early in the season and there aren't many places the bees can forage, you can put a feeder on top of the super. Now replace the inner and outer covers and you are finished. Check your new colony within a few days.

Bait hives

Bait hives are an easy and inexpensive way of obtaining honey bees to start a new colony. Captured swarms can be used for increasing colony numbers, starting a bee hobby or bolstering weak colonies. All you need is an empty hive to act as a bait station to encourage the bees. When bees require a new home they send out scout bees to find good places and to assess their value. These scouts report back to the main group of bees and feed them information about the sites they have found. They do this by using the waggle dance explained earlier in the book (see pages 20–21).

Your job is to encourage the scout bees into thinking that your hive is the best home for the swarm. This is done by baiting the hives you want to be colonized. Beeswax is considered to be the best bait to attract swarms and starter strips on the top bars may be enough to do this. If this does not work you will need to spread more beeswax around the hive.

A hive that has already contained bees and harvested honey will be more attractive than a brand new one. Bees also find the scent of a brood very appealing, so a top bar with a small section of comb containing brood – even if it is dead – will make an excellent lure. Obviously you will need to make sure that there is no disease present in the section of comb. Do not leave the comb for more than a week as it will start to

Left: *Scout bees are sent out by the hive to seek potential new homes and pollen sources.*

attract wax moths and will no longer be attractive to the bees. Make sure you have enough time to visit the hive on a regular basis to see if your bait has done the trick. Although beekeepers try every trick in the book to avoid swarming, it can be an asset to keep a bait hive in an apiary just in case something goes wrong. Because bait hives are usually made of old equipment, they are not ideal to be used as a permanent home for a colony, but it is easy to transfer them to another hive by placing a super of combs on top.

The advantage of bait hives is that they are very inexpensive. You simply use old equipment and just sit and wait.

Above: *A simple, box-style bait hive set up in a residential garden.*

USING OTHER BAITS

Beekeepers in different parts of the country have come up with their own type of lures. Honey is not a useful bait, as bees just rob it very quickly, although you could rub a little above the top bars where the bees can smell it but cannot actually reach it. It does also mean that the bees will quickly learn the location of the hive and scouts may return to it later.

You can also use a pheromone lure (nasonov pheromone), which is more expensive than the previous method but is very attractive to bees. Commercial swarm lures are also available.

Nasonov pheromone lure

This is named after the nasonov gland found on the honey bee. It is a pheromone used by bees as a signal to other bees to say where they are, or that they have found a new nest. This lure duplicates the scent that scout bees release when they find a good home for their swarm.

To use this lure you need to place a small needle hole in the lid to allow sufficient pheromone to waft out of the hive. This hole can be sealed up when not in use with a drop of beeswax. Position the bait into a swarm trap which is wired about 50 mm above the entrance hole.

Researchers have discovered that the following factors make the bait hive more attractive to swarms:

- Swarms do not seem to take any preference over the shape of the cavity or the entrance, but they do prefer those that contain nasonov pheromone.
  Traps have been found to be more effective if they are placed about 4.5 metres above the ground.
- Lures that are used in combination with old combs and hive residue odours such as propolis appear more attractive to the scout bee.
- Swarms prefer an entrance hole approximately 25 mm in diameter which is located towards the floor of the hive.
- Swarms seem to prefer those hives that face south.
- Scout bees prefer sites where there are landmarks or points of interest to use as navigation guides.

It is important to keep a check on empty hives during the colonizing season as you need to make sure that they have not become homes for unwanted residents such as ants, beetles, rats or spiders, to name but a few. Despite all your best efforts you might find that you haven't attracted a swarm to your bait hive. Don't give up – you can always try again next season.

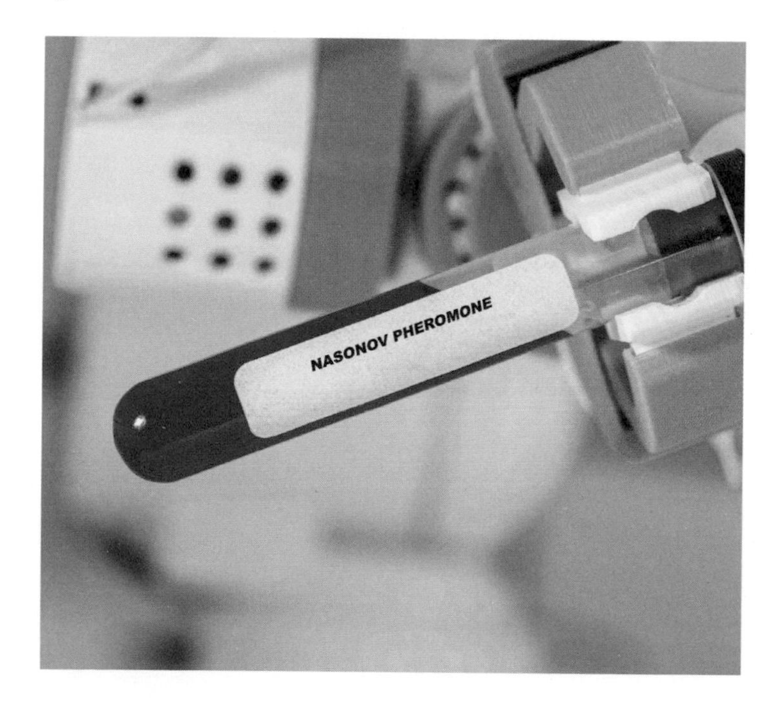

Left: *The Nasonov pheromone lure.*

CHAPTER THREE

THE BEEKEEPER'S YEAR

Understanding the seasonal management of a colony is of the utmost importance. As you may not remember exactly what takes place from month to month, it is worth keeping a journal of actions you have taken, so that you can replicate the successful ones the following year.

Below: *Early spring plants including snowdrops and winter aconites.*

The beekeeper's year starts in spring, and this is the ideal time to commence your new hobby. Proper hive management is the key to successful honey production. If you follow the advice given in this book you will be able to watch your small nucleus of bees grow into a thriving colony, and your experience will grow month by month. Colonies of bees actually need very little management as long as they are given the right conditions, but there is still quite a long list of jobs to do during the year. Try to resist the temptation to open up hives until a really suitable day arrives – around 14°C/57°F – to avoid chilling the brood. If you can go out comfortably in short sleeves and your hat doesn't get blown off in the wind, then you can consider this a suitable day. On a warm spring day, your bees will start to make cleansing flights. This is when you might see streaks of yellow on your washing! Although there will not be many plants in flower to encourage foraging in early spring, the following can coax the bees into early activity:

- Winter aconite (*Eranthis hyemalis*)
- Snowdrop (*Galanthus nivalis*)
- Crocus (*Crocus spp.*)
- Gorse (*Ulex europaeus*)
- Hazel (*Corylus avellana*)
- Willow (*Salix caprea*)
- Yew (*Taxus baccata*)

Spring

Your new bees have arrived and you have left them alone for the first week to allow the colony to settle into their new surroundings. Now you must make sure they have plenty of room to expand and that they are disease- and problem-free. You need to learn to recognize their needs. You should aim to inspect your hives at least once a month, but ideally it should be more often until winter dormancy takes over.

THE FIRST SPRING INSPECTION

The purpose of the first early spring inspection is to see what is going on since you introduced your bees. So what signs are you looking for? Remember to keep the first inspection fairly brief, as you do not want to risk chilling the new brood.

☑ Check that you can see the queen and that she is laying eggs.
☑ Check that there are brood of all stages present in the cells.
☑ Check for any diseases and pests.
☑ Check the general cleanliness of the floor.
☑ Check that the colony has a sufficient supply of food.
☑ Check that the colony has started to increase and count the number of frames now covered.
☑ Check that the bees have enough room in the hive.

You will also need to be checking to see if the colony is about to swarm and for any queen cells in the hive. To the beginner this might sound a rather long list of jobs to do, but you will very quickly learn how to recognize any problems and deal with them appropriately. Before talking about opening the hive, it is worth inspecting the outside for any telltale signs.

LOOKING AT THE HIVE

You can tell a lot about a hive from checking the entrance to it. If you see bees crawling in front of the hive and excreta deposits near the entrance, this could indicate the presence of a disease such as nosema. (See Chapter Five: Hive Problems, Pests and Diseases for full details.) If there is a pile of dead bees at the entrance then it could indicate some internal problem with the hive. Examine the inside of the hive to find out whether the bees have fallen prey to varroa mites or some other nasty disease. If there are bees fighting around the entrance, then this indicates that robbing is taking place (see page 99). If you see a lot of bees coming out of the hive in a swirling mass then it indicates that a swarm is about to

Above: *Making an inspection of the hive is important to check on the health and vitality of your bees.*

emerge. If you see dead larvae being thrown out of the hive but not being carried away, this indicates that there is not sufficient food for the colony to survive. On a better note, however, if you see bees carrying pollen into the hive, this indicates a healthy colony. Finally, if you see a regular column of ants going into the hive and there are very few bees around, then it would indicate that the hive is empty. Put your ear to the side of the hive and give it a hard knock. If you don't hear a loud buzzing, then you need to check your hive. It is a good idea to check the entrance to your hives every time you pass them, even if you don't intend to make a full maintenance inspection. It doesn't take very long and it can alert you to any possible problems.

HOW TO CARRY OUT A FULL INSPECTION

This inspection is aimed at the beekeeper with a new colony that is about one week old, but as your brood expands you should still follow the same pattern. Light your smoker, just in case you need it, put on your protective clothing and make sure you have your hive tool handy. Now follow this useful list.

1 Open the lid

Approaching the beehive from the side, not the front, gently lift the lid using your hive tool if necessary. Place the lid on the ground upside down, as this gives you somewhere to place the other boxes later on. This doesn't apply, of course, if you have chosen the type of beehive with a gabled lid. If you are feeling a little nervous about entering your hive, then you can puff a little smoke over the top bars, although this is not usually necessary with a new colony as they should be fairly passive. If you do use smoke you will see the bees that are nearest the top quickly disappear further into the hive and immediately start to gorge on their honey stores.

2 Check for eggs

If you have a feeder frame at the end of your box, it is a good idea to remove it at this stage to give yourself more room. Now remove one of the end frames, gently separating the frames with your hive tool. Make sure your frames are not stuck to another one before you attempt to lift them out. It is the centre frames that will hold the brood, and it is these ones that you want to inspect.

Remove the frame with the brood on it and hold it up to the light. You are trying to see if there are any eggs in the base of each cell. They are not easy to spot for the untrained eye, as they are tiny, white objects which look like minute grains of rice. If you find a single egg in each cell, then this means you have a healthy, laying queen. If you see more than one egg in each cell and they are not at the bottom where they should be, this indicates that you have workers laying. The chapter on Problem Solving will give you more details, but it is fair to say you need advice.

If you cannot see any eggs at all in the cells, then you either have no queen or a queen that is not laying and she will need to be replaced.

3 Finding the queen

The next task on your list is to try to find the queen. She is recognizable by her long abdomen and shorter wings and she will be moving much slower than the other bees. You should be able to see her on a frame of brood, but this is not always the case. If you can't find her and you know you have eggs, then don't panic, the chances are that she is very good at hiding. You may have been lucky enough to have bought a colony with a marked queen, in which case she will be much easier to spot. If you do see her, just have a quick look to see if she appears healthy and then gently lower the frame back into the brood box.

Even experienced beekeepers have problems trying to find the queen, so you may not spot her on your first inspection. You will learn on page 80 exactly how to mark your queen, which will make your job much easier, especially in a large colony.

4 Studying the brood nest

Learning to understand the condition of your brood nest is one of the best ways of determining the health of your bee colony. What you are looking for are signs of healthy brood in all their stages – eggs, young larvae, larvae and larvae in capped brood cells.

The presence of brood in these stages indicates that the colony is functioning correctly. The eggs lie at the base of each cell in the brood nest, in a bed of milky brood food. After six days the larvae should have increased in size to almost fill the base of the cell. Healthy larvae are pearly-white in colour and lie in a distinct 'C' shape, with the head and tail curled towards one another. When the larvae are nine days old, the cell opening is sealed by adult bees with a cap of wax and the adult bee completes its metamorphosis. The wax cappings on healthy worker brood can vary in colour from very light to dark brown. A good brood pattern with very few empty cells, suggests that the queen is laying well and that nearly all the larvae are developing normally.

Below: *Carefully inspect the frames to check on the brood, honey stores and the number of bees.*

Make sure that the cappings are slightly convex with no holes and no sunken parts. This is easy to see if you only have a small colony, and this is why starting with a nuc or package is ideal for the beginner. If there is no evidence of eggs or a young brood and you are unable to find the queen after a thorough check, then the colony will need to be united with another colony or nucleus.

5 Inspecting the food stores

Now that you know the state of your brood and that you have a healthy, laying queen, you need to make sure the bees have enough food stores to survive. They need stores of both honey and pollen, or at least a feeder full of sugar syrup. The stores will be positioned in an arc around the brood area, so checking for the presence of honey and pollen is easy. If there are no full cells, then you must do something about it and feed the colony immediately.

6 Spring feeding

It is not known exactly how much honey and pollen a colony of honey bees needs to sustain itself throughout the year, but it is estimated that a colony should have at least 6 to 9 kilograms of honey in reserve at all times. It is easy for a colony to suffer from starvation when there are too few plants in flower in the area, and their first reaction is to remove the larvae and discard them in front of the hive. The queen will continue to lay her eggs, but these larvae will be discarded, too. The overall effect of this starvation is that the colony will become weaker, perhaps even resulting in death. This is why the beekeeper needs to be alert as to the food reserves in his colonies.

Many beekeepers will save combs of honey from the autumn to allow for spring feeding. This is probably the best and easiest way to feed the colony as this will not usually stimulate robbing. Just be careful when you are feeding sugar syrup that you don't spill any drops on the ground around the apiary, as this can encourage robbing.

Lack of pollen is one of the biggest causes of a colony failure. Bees need pollen to provide them with protein and if they have a shortage, then it is up to the beekeeper to feed them with a pollen substitute. Either consult your local beekeeping association or follow the recipes in the section on surviving the winter (see pages 96–108).

7 Checking for adequate space

Having ascertained that your queen is laying and that your colony is increasing satisfactorily, your next task is to check that the bees have enough space. There needs to be sufficient empty cells for the queen to

Above: *Wax moth larva on the honeycomb. An infestation of wax moths can reduce the comb to a desiccated husk.*

continue laying, remembering that she can lay as many as 1,000 to 2,000 eggs each day. There also needs to be enough empty cells for the bees to store food. If you have installed a small nuc, then this should not be a problem initially, but as the brood expands you may need to add another brood box to your hive. If the bees start to feel overcrowded then you will run the risk of swarming. If you allow this to happen you are going to lose around half of your workers. They will quite simply move to someone else's hive and provide them with honey.

8 Watching for signs of disease

For a new beekeeper, spotting signs of disease can be difficult. However, starting with a new, young colony will help you to recognize what a healthy colony looks like and consequently if you spot anything that doesn't look right you will know you have a problem. As a beginner it is probably a good idea to find a book with pictures of possible diseases so that you can take it with you to the hive for easy identification. Always make this one of your priorities when making an inspection of your hive, so that you can eradicate any problems before they take over the whole colony.

9 Looking for wax moths

A healthy bee colony will be able to control wax moths, but it is still a good idea to kill any that you see roaming around your hive. The wax moth is grey in appearance and you may see both large and small varieties. The larvae of the wax moth can be voracious and if left to build up in numbers can wreak massive damage in a very short time, so this is another important part of regular inspections.

10 Floor inspection

The floor of the hive will tell you a lot about what is happening in your colony. To inspect the floor, gently lift the brood box and place it carefully on the upturned lid – any jerks or sudden movements could dislodge the queen. Check to see if the floor is clean and free from debris. If there is a build-up, use your hive tool to scrape the floor clean, or alternatively swap it for a new floor and clean the old one later when you have more time.

If the floor is very dirty, this could indicate that you have a problem, so look around for any warning signs. If you have a solid floor and you notice a build-up of water, then this indicates that you have not tilted the hive enough to allow water to run out of the entrance.

11 Bee inspection

Now is a good time to lift out some of the frames, one at a time, and observe what the bees are doing. To the novice their actions will appear to be chaotic, but if you look carefully you will see it is organized chaos. You should be able to see a bee with its pollen baskets full, moving towards the pollen storage area. You might see another bee dancing, communicating with other members of the colony where to find the best food sources. Hopefully you will see adult bees cutting their way out of the wax cappings and slowly emerging and, if all goes well, you will see the queen too. Always remember to replace the frames in exactly the same order as you took them out, so as not to cause confusion to the colony.

Below: *Inspecting the frames to check on the activity of the bees.*

12 Clean equipment and good combs

When you have had your hives for more than one season you will need to check for wear and tear. Replace worn and warped equipment, broken combs, combs that have too much drone comb, inner cover and any other part of the hive that is showing signs of deterioration. Spring is an ideal time to do this maintenance work as there are fewer bees in the hive.

During the course of the year bees collect and bring a lot of propolis back to the hive. An accumulation of this sticky substance can make the removal of frames extremely difficult, so make sure you don't allow an excess to build up.

13 Reassembling the hive

Once you are satisfied that everything is going well with your new colony, you can reassemble your hive. If you have more than one hive in your apiary, move to the next one and repeat the whole process.

Having established that your colony is thriving and healthy and you haven't spotted anything untoward, then you can rest in peace until your next inspection. However, there is one final item on the list of things to do and that is to take preventative measures against varroa mite. Although some strains of bees are able to combat this invasive mite, the western honey bee, *Apis mellifera*, has not evolved a defence mechanism. The varroa mite has managed to invade virtually everywhere that bees are kept and an infestation will quickly destroy your colony. During the spring, when your colonies are growing fast, so are the varroa mites. There are many treatments and many different ways of administering them, so it is advisable to find out from your local beekeeping association how they handle the problem and follow their advice.

SPRING SWARMING

You might feel that swarming crops up quite frequently when reading anything about honey bees. This is because preventing swarming is the most difficult task that a beekeeper faces. Swarm control is an important part of spring management and is often the most frustrating. Having concentrated in the early spring on building the colony up to a sustainable size, the beekeeper is suddenly faced with the other extreme – the hive has been overpopulated and is ready to swarm. Swarming is a completely natural process and without it, the species would not survive. It is up to the individual beekeeper to recognize the signs that a colony is about to swarm and to do everything they can to prevent it.

As a result of certain conditions within the hive, the colony will start to raise new queens. You will notice that worker bees have started to build small queen cups on the comb, mostly near the comb's outer edges and especially along the bottom. These don't immediately indicate that the bees are about to swarm, but when the conditions are right, the queen will be directed towards these cups to lay one or more eggs in them. As soon as this happens, it is time to take preventative measures.

If you leave these cells untouched, a virgin queen will emerge after approximately 16 days to take over the colony. The old queen will have left in the swarm. This new virgin queen will go to the other queen cells and sting her rivals to death.

Above: *Keep an eye on whether the colony is raising new queen cells. Their presence can indicate that all is not well in the hive and may be a precursor to a swarm.*

SWARM PREVENTION AND CONTROL

The main thing to remember is that if you do allow your bees to swarm, then half your honey-collecting workers will go with the swarm. It will take a long time before your hive is up and running again, because the bees will have to wait for the new virgin queen to mate and start producing a new brood.

Swarm prevention is taking steps to deter or prevent the construction of queen cells within a colony and managing the colonies so that the swarming impulse does not arise. Swarm control is something completely different. This means that you allow the bees to swarm but it is under your control so that you manage to retain the bees.

How to stop your bees from swarming

You may have read about many different ways to control swarming, including clipping the wings of the queen bee and finding and destroying queen cells. These methods are used but are probably not advisable for the beginner. If you use the following steps as guidelines you should be able to keep swarming to a minimum. Good swarm-prevention methods are those that involve the least interference on the daily workings of your colony. The best and most assured way of preventing swarming is to purchase a strain of bees that has a lower tendency to swarm, but of course this is not always possible.

Above: *A new queen held in an introduction cage, ready to be placed into the hive.*

1 Requeening
This should be carried out annually or at least every two years. This is an ideal method for a beekeeper who only has a few hives to manage. Queens who are less than one year old with a good supply of pheromone are less likely to swarm than a queen in her second year. Once a queen starts to age her pheromone levels drop and the workers are unable to smell her, or think that she is not present in the hive.

2 Adding supers
This method is simple and involves putting supers on top of the brood boxes in time for the honey flow. The first box should be filled with comb, especially if it is early in the season. Adding supers in time is not only essential for the preparation stage of storing honey, but it also limits swarming as it gives the bees more room in the hive.

3 Equalize your colonies
You can try to equalize the strength of your colonies in two ways. You can move frames from a strong colony to a weaker one, or you can swap the position of the weak and strong hives. This latter method seems to work quite well on hives that are showing signs of swarming. Of course before swapping bees between colonies you will need to make sure that they are free from disease, otherwise you risk spreading it. If you manage to even out the number of bees in your hives, not only will you have stronger colonies, but they will 'perform' at about the same time, making your life a lot easier.

4 Good ventilation
Good ventilation is essential if you are to stop swarming. Make sure that entrances are appropriate for the time of year, that the lid of the hive is painted white and that you use a stainless-steel mesh floor. If you live in a hot climate, then shade boards over the entrance can help control the internal temperature.

5 Reversing method
The reversing method is the simple procedure of swapping the positions of the upper and lower brood boxes. If you only have one brood box in place, you can place a second brood box on top of the first. If you are placing a new box on top then you will need to fill it with frames of

empty comb, with one frame of capped brood from the existing brood box in the middle position. This will give the colony more room and reduce the risk of swarming. Try to make sure that you do this reversal before the queen cells are started. After about two weeks, reverse the bees again if the bees have already moved up and keep doing this until the end of the season.

WAYS OF CONTROLLING A SWARM

The chances are that you will miss the signs of swarming during your first season as a beekeeper, but don't panic. If you do find there are queen cells already formed, then you can at least take steps to make sure the inevitable swarm will stay in your apiary. The first thing to do is to check to see if the colony has already swarmed. Even if you are too late and they have left the hive, you can still carry out the following:

Simple swarm control

This procedure can be carried out if your inspection has shown that queen cells and normal brood conditions exist. You will, however, need a little extra equipment to perform this swarm control:

- An extra brood box with frames and foundation or comb
- A cover board
- A roof
- Enough space around the original hive. Now take the following steps:

Left: *Use a smoker to subdue the bees if you feel that they may be about to swarm.*

1 Subdue the bees using a little smoke and remove the roof, cover board and super from the original hive. Replace the cover board directly on the brood box to settle the bees. Now move the original brood box to one side of the original location – ideally about 1 metre apart.
2 Position the extra floor and extra brood box with frames and foundation on the site of the original hive, taking out the two centre frames, keeping them for later use.
3 Carefully go through the frames in the original hive until you find the queen. Remove the frame with the queen on it, check for queen cells and remove any found. Place the frame with the queen and young bees into the extra brood box on the original site.
4 Take another frame from the original hive that has sealed brood and a good number of young bees, check it for queen cells and remove any found. Place this frame in the remaining space in the extra brood box.
5 Now place the original queen excluder, super, cover board and roof on to enclose this 'artificial swarm'.
6 Next move all the frames in the original hive together and close the gaps at the ends with the two frames you put aside earlier. Close up this hive with a cover board and roof and either relocate it or turn the entire hive 180 degrees with the entrance facing in the opposite direction to the swarm hive.

Swarm control summary

The queen, along with young bees and sealed brood have been taken from the original hive and placed in a new location. The remaining original frames, brood and young bees will act as if the hive has just swarmed, but to a new location. It may have queen cells, but if not, the young bees will quickly create new emergency queen cells where needed.

The adult bees will now return to the original location and boost the numbers in the artificial swarm, giving you two thriving colonies.

If you did not find any queen cells in the original hive, wait about five days and then carry out a check for the presence of emergency queen cells. Select the cell with the most advanced larvae and destroy the remainder. Alternatively, if there were queen cells in the original hive, select the most advanced and destroy the others. Then leave both hives alone and let nature take its course.

Now you have two choices. You can either allow both hives to develop into full colonies over the summer months; or, alternatively, get rid of one of the queens as autumn approaches and combine the two hives to produce one strong unit for the winter months.

The two-queen system

Before moving on to summer tasks, it is worth mentioning that there are ways of increasing your harvest, and one is to use two queens in one hive. This gives you the added advantage of:

- Higher yields using less equipment
- Control of swarming
- Simple requeening each year
- Strong colonies

The main drawback to this method is that the two queens must be kept apart, otherwise they will fight and you will end up losing one anyway. In addition, you have to be able to predict the nectar flow fairly accurately, and know what plants are growing in your area.

This system has not gained much in popularity, and yet it can produce impressive yields and there are now special beehives available to operate it.

If you want to try the two-queen method, there are a number of rules you should be aware of. Follow these simple tips, and you should find you have plenty of honey at the end of the season:

1. Only use a colony that is really strong.
2. Divide the colony a couple of months before the expected start of honey flow.
3. Place the old queen and young brood (or uncapped cells), along with about half the bees, in the bottom chamber.
4. Cover this chamber with a division board.
5. Place a new queen with some capped brood and the other half of the bees in the upper chamber.
6. Above this, place an empty brood chamber which contains drawn combs, if you have any spare.
7. Keep reversing the brood chambers, as in the swarm-prevention techniques.
8. After about two weeks replace the division board with a queen excluder.
9. As the flow starts to come, put on as many supers as required.
10. About a month before the end of the flow, remove the queen excluder so that both colonies unite. Only the new, stronger queen will survive as the old queen will be killed. The colony can then spend winter with the new queen.

Marking the queen

You have already learned that the queen is the most vital element to your hive – without her the colony would perish. The queen is usually quite difficult to find and marking is an easy way of keeping track of where she is and exactly how old she is. It is also handy if you find an unmarked queen in your hive, because you will know that either swarming or supersedure has taken place. There are, of course, disadvantages as well. For example, you could damage the queen while you are marking her, which could lead to rejection by the colony. You may also get so used to seeing a queen with a dot on her thorax that your eye misses unmarked queens. However, the advantages certainly outweigh the disadvantages. Queens are marked using the International Marking Code:

YEAR	COLOUR
0 or 5	Blue
1 or 6	White
2 or 7	Yellow
3 or 8	Red
4 or 9	Green

METHOD

It is not a good idea to touch the queen by hand as you could either damage her, or your scent could make her rejected by the colony. You can buy special cages for the purpose of catching the queen, which you hold over her on the comb. It doesn't matter if you pick up a few workers at the same time as it has slots in the cage which are large enough to allow them to escape. Take the cage into a shed or small enclosed area so that if the queen does escape you will not lose her.

You can either mark the queen in this cage or put her into a special marking cage which has a sponge plunger and a metal screen at one end. Place the queen inside the cage and gently push the plunger until she is pushed up against the mesh, making sure of course that she is the right way up. She is then in the ideal position to place a spot of marking fluid on her thorax. Make sure you are wearing your glasses if your eyesight is failing, or alternatively have a small magnifying glass to hand so that you can see exactly where you are marking her. You can buy special marking fluid, marking pens or you can just buy a pot of Tippex™ correcting fluid which works equally well, but of course this only comes in white.

Leave the queen in the cage for a few more minutes to make sure that the fluid is dry. Now it is time to return her to the hive. Make sure that

Right: *Marking the queen helps the beekeeper to spot her among a sea of workers.*

she is over the hive and very close to the brood frame before pulling out the plunger. You may find she does not drop out immediately, so a gentle shake will soon encourage her back into the hive.

It is a good idea to keep a record of any queen marking with some sort of numbering system. Although marking the queen is not essential, it is a good practice to get into and makes life much easier when you are doing your inspections.

The following are the obvious advantages to marking the queen:

- A queen which has been both marked and recorded will be 'dated' and so it is always possible to determine her age by referring to your notebook or record card, whichever system you choose to use.
- She can be easily and quickly found at any time.
- It is also possible to ascertain whether she has been superseded, or if she has attempted to leave with a swarm, in which case she is lost.
- If you decide to breed your own queens, one of the greatest advantages of marking is the benefit of knowing exactly how long they live. This is especially important because it means you will be able to ascertain which bee strains live longer than others. This means they can continue to lay vigorously for much longer.

Spring checklist

This section looks at spring month by month so that you can make a checklist to see if you have carried out all the necessary tasks for the forthcoming season. In many ways the management in spring will set the scene for the rest of the year.

MARCH

☑ This month can still be cold and wet and the bees may only fly occasionally. They will be developing a large brood nest and many of the old winter bees will be dying. This is the most critical month for maintaining the colony and many colonies starve during March because they were not adequately fed in the previous autumn.
☑ Check that there are sufficient stores by hefting. This involves the estimation of hive weight by lifting the box.
☑ Start a little feeding with light sugar syrup to stimulate the colony to build.
☑ Remove mouse guards and other protective equipment (for woodpeckers etc.)
☑ Check the entrance to the hives to make sure that the bees are flying and busy. If there is a mild spell the bees will become more active, taking advantage of early flowers such as the aconite, snowdrop, crocus and butterbur, so look for yellow pollen loads on the bees' hind legs.
☑ If there is a very mild day you could make a superficial examination of the brood chamber.

APRIL

☑ Willow, flowering currant and some ornamental cherries provide the first sizeable flow of pollen and nectar. Brood-rearing increases and bees are able to take regular flights.
☑ This is the first month when it is necessary to inspect the colony in detail. However, only check the brood nest if the weather is warm, calm and sunny (say above 18°C) as you do not want to chill the brood.
☑ On a warm day, spring-clean the hives. It's important to clean all hive parts to prevent the spread of any diseases that might be lurking in the build-up of the previous year's wax and propolis.
☑ This is the time to mark any unmarked queens.
☑ There are five main things to look for this month:

- Is there a laying queen?
- Is the brood pattern even?
- Are there enough stores until the next inspection?
- Is there enough space for expansion and storing of pollen and nectar?
- Is there any sign of disease?

☑ Move any old and dirty frames to the side of the colony for removal later. Do not, however, disturb the overall pattern of the brood nest.
☑ Occasionally in good weather or when the hive is near a spring flowering crop, there will be a surplus of honey and it will be possible to remove some supers of honey to give a spring crop.

☑ After checking for disease, make your first treatment for varroa mite. Winter diseases to look out for are acarine and nosema which will be indicated by crawling bees and dysentery.

PROVIDING SPRING FLOWERS

Although honey bees will normally forage some distance from the site of the apiary, it is still a good idea to provide nectar- and pollen-rich flowers throughout the year. The more flowers you have, the more attractive your garden will be to bees, so you can never have too many flowers. Even if they do not attract your own bees, it will be a welcome supply to wild bees, bumble bees and other nectar-collecting insects. The major nectar and pollen sources are well-known to the beekeeping fraternity, but if you are not sure then there are many books on the subject. Here is a selection of good spring and summer flowers, both cultivated and wild, to have in or close to your garden.

Below: *Making the most of spring blossom – in this case an almond flower to collect a variety of pollen*

Alyssum
Apple
Bluebell
Bottlebrush
Broom
Bugle
Cherry
Cotoneaster
Erica carnea (heather)
Fennel
Flowering currant
Hawthorn
Lavenders
Lungwort (*Pulmonaria*)
Marigold
Pear
Plum
Rosemary
Sage
Salvias
Zinnia

Beekeeping records

Because keeping records is a vital part of beekeeping, this has been included in the spring section as this is the time to form some sort of system. Everyone has a different way of keeping records, but this is intended to give you an idea of how to plan your notebook or card system.

Hive records are a convenient way of showing the state of the colony each time it is inspected or manipulated. Some beekeepers prefer to keep their record cards next to their hives, in which case you will need to make sure they are kept dry and away from the bees as they will chew the card and your records will be lost. Some people prefer to keep notes similar to those in a diary, while others prefer to use a computer. If record keeping is to be successful, it is important that the records hold the required information and that they are quick and easy to complete.

When planning your records, a good starting point is the five main questions that will crop up again and again.

- Room – has the colony got enough?
- Queen – is the queen present and are there any signs of swarm preparation?
- Development – how many frames of brood are there?
- Disease – are there any signs?
- Stores – has the colony enough stores to survive the next inspection?

Below: *Keeping effective and detailed records helps you to monitor the progress of your bees throughout the year.*

If you make sure your records answer these five questions at each inspection, then it is possible to keep track of the condition of your colony. It also helps you to compare the development between colonies and to identify those colonies which are better or worse than others.

By adding additional columns to your chart, it is possible to record other aspects of the colony, for example, temper, honey crop, weather conditions and so on. It is up to you exactly how complex or simple you would like your records to be.

The chart on the following page is intended to act as a guideline to the new beekeeper who, as yet, has not worked out a system for keeping records. There are no specific rules or specifications in setting up your records, but choose a format that you can understand and that is easy to use.

APIARY No.	COLONY No.
Date	4 March
Q	x
QC	x
Swarm	N
Brood	✓
Stores	5
Room	5
Health	✓
Varroa	1
Temper	10
Feed	1 LS
Supers	0
Weather	S, 15
Notes	Removed mouse guard. Cut grass under hive

KEY TO SYMBOLS:

Date	Date of the inspection
Q	Presence of queen
QC	Presence of queen cells x = none seen 10X = 10 seen but all removed 2L = 2 seen but left alone
Swarm	Are there signs of swarming? Y = yes; N = no
Brood	Condition of brood e = eggs seen ✓ = brood pattern OK 3 = brood covering 3 frames x = no brood
Stores	Quantity of stores available 10 = 10 super frames available
Room	Available space for queen to lay 5 = 5 brood frames available
Health	State of brood and adult bees ✓ = all OK CB? – chalkbrood possibly EFB? = European foulbrood possibly, etc.
Varroa	The number of varroa mites in a colony l, m, h = low, medium, high
Temper	Docility of the colony 10 = nice calm bees 8 = agitated bees 6 = bees sting 4 = bees that follow too much
Feed	How much feed given 2 LS = 2 litres of light syrup 1 HS = 1 litre of heavy syrup
Supers	How many supers removed or added +1 = 1 super added –5 = 5 frames removed, etc.
Weather	Temperature and cloud cover c = cloudy s = sunny r = rainy f = fair
Notes	Anything of interest that needs to be added.

Left: *You could use this chart as a guideline for your records or set up your own system.*

Summer and autumn

Summer and autumn are what are referred to as the 'active season' in the beekeeper's year. By the onset of summer you will have established your hive or hives in either a rural or urban site and got used to inspecting them every ten days or so. You will probably have carried out swarm prevention and control measures and you know that your colony has room for expansion.

From now on your aim is to assist your colony in any way possible to build up to their maximum strength before the main flow starts. 'Honey flow' is a beekeeping term denoting the time when one or more of the major nectar sources are in bloom and when the bees therefore produce and store surplus honey.

High bee numbers are important – a colony of around 60,000 bees can collect as much as 50 kg of honey during the season, whereas two smaller ones, even though they have the same number of bees in total, will only collect 45 kg.

Below: *The beekeeper's priorities change as the year progresses; in summer you will be focusing on honey production.*

The swarming season is coming to an end, so what do you need to do now? With the onset of the primary honey flow, inspect your colonies and if you find that some are weaker than others, now is the time to unite the colonies. You should have already noticed an increase in foraging activity and the combs will be showing extra nectar content. This is the time of year when you need to make sure that your bees have enough room to store honey, and this should become your main priority.

HONEY-FLOW MANAGEMENT

If bees are provided with drawn comb, it has been proved that bees are encouraged to store more honey than they require. As long as there is a sufficient honey flow and you have enough bees, then they will keep storing more and more honey. There is a lot of confusion as to how many supers to add to a hive. The fact is there is no general rule and this is perhaps one of the great fascinations about beekeeping.

Management of the hive during this period is crucial as it is the time when the bees need room to store honey and you need to get the timing right. The timing will depend entirely on where you live and the primary plant nectar sources. Once your nectar sources start blooming, the honey flow will begin so follow these steps:

1. Ten days to two weeks before the major nectar-producing plants begin to bloom, add a honey super on top of the brood chamber.
2. To keep the queen out of the honey super so that she won't lay eggs you can take two preventative methods. You can either use a queen excluder or you can use two or three frames of plain foundation in the middle of the honey super with drawn comb frames on the outside. A queen will always come up to the middle of the super when moving up and if she sees the plain foundation she will not lay eggs there.
3. Keep an eye on how full the honey super gets and add another one if it starts to get full. It is far better to have too much space than too little, because if the bees run out of room there is still a risk of them swarming. It really depends on how strong your colony is and how much honey you think they can produce. It's better to give them too much, rather than too little, room.
4. When the main honey flow has finished – this will depend entirely on where you live – you can take off the honey supers and start to harvest the honey. This could be as early as July or as late as August/ September depending on the types of plants in your area and when they bloom.

ADDING FOUNDATION AND MAKING NEW COMBS

It is good beekeeping practice to make a few new combs every year. As combs age, the cells become smaller with each cycle of brood because of the accumulated cocoons. They are also prone to wear and tear, and old combs often contain too many drone cells. Combs that have not been properly wired can sag and stretch the cells.

The best time to draw foundation or to make new combs is during the honey flow. Although bees will draw foundation at any time during the active season, they do their best job during the flow. Place the new frames with foundation just above the brood nest. The ideal situation is to place six new frames with foundation in the middle of the second super with an excluder on top and a queen below. The outside combs – in addition to the six new combs – should be drawn comb.

Above: Top: *Adding a new frame with fresh wax foundation to encourage the bees to make new combs.* Below: *Uncapping the cells allows the honey to flow out of the comb.*

Another way to draw foundation is to place four new combs, spaced alternately between five drawn combs, or to place an entire super of foundation immediately above the super in which the bees are storing nectar at that time. Some beekeepers make a habit of adding one or two new combs to each super as they are added during the honey flow.

The reason that new foundation should be added during the honey flow is that the bees have a tendency to chew the edges of the wax foundation if nectar is readily available. This is why sections for honeycomb production are not added until the honey flow is underway.

HELPING YOUR BEES STORE HONEY

You will be amazed just how quickly a healthy colony builds up honey stores when the conditions are right. When this happens, there are several ways of giving them a helping hand. As explained previously, you can provide supers full of comb.

You can also help your bees by making entrances in the supers in the form of a drilled hole 1 cm in diameter. This means the workers can go straight to the super without having to work their way through the brood areas when their sacs are full of pollen.

REMOVING THE CROP

As soon as the honey flow is finished and the combs are full to the brim with new honey, they should be removed and the honey extracted.

The reason that extraction should not be delayed, is that certain types of honey will crystallize in the comb and once they do, it is extremely difficult to remove the honey. Bees that have foraged on asters, for example, will produce a honey that can crystallize within 10 to 20 days of being stored in the combs. For full details on harvesting your crop go to the section on Extracting the honey, pages 112–119.

AUTUMN MANAGEMENT

Having stressed the importance of spring management, the autumn months are just as vital as mismanagement at this time could mean losing the colony over the winter. This period is often overlooked after the rush to take in the honey crop, so this is intended as a reminder of things to do in late summer and early autumn, to prepare for a successful wintering.

Below: *The end of summer does not end the beekeeper's to-do list! Autumn management is key to the continued success of the colony.*

You could encounter a few problems if there is a later honey flow, which could possibly happen during an Indian summer or just because the autumn is unusually mild. If you have already reduced the hive in preparation for winter, left to their own devices the bees will fill the brood area with honey, cutting down on the amount of brood area available. There is always a danger of robbing and bees will not hesitate to rob out weak hives, creating many other problems besides the brood area being clogged with honey. It is imperative that the beekeeper keeps a close eye on whether it is necessary to feed sugar syrup, or whether the bees are still finding their own supplies of nectar.

Although you might think it is a good idea to have a build up of honey stores before going into winter, the loss of brood area will drastically reduce the number of young bees. This, in effect, forces the hive to winter older bees, reducing the population of young bees who are essential for feeding brood early in the spring. For this reason it is not a good idea to remove the supers too early. It is far better to leave one in place until you have had the first frost. Bees prefer to move their stores upwards rather than fill any brood space with permanent honey.

If you do allow the numbers to diminish, there will be insufficient bees to keep the cluster warm enough to prevent super chill in the depth of winter and this in itself is a recipe for disaster.

Make your autumn management start at the end of summer. If your inspection shows there is an excess of pollen clogging the centre frames, they should be removed and replaced with empty drawn comb. Some hives will gather excessive amounts of pollen far beyond their normal requirement stored in the bottom box. This in effect stops the queen from moving down into the bottom box, preventing the ideal position of the brood area going into winter. Remember to try to keep the centre frames open by replacing them with a few empty ones. This will prevent a clogged hive.

Autumn is an important time of year to manage hives for the prevention of spring outbreaks of nosema, European foulbrood and chalkbrood disease. Hives showing signs of disease in autumn require extra attention during the winter to prevent a major problem in early spring. Placing hives in a warm, sunny location during winter helps in reducing the levels of nosema.

This is also an ideal time of year to requeen a colony. A colony led by a young queen coming into spring will have a larger brood area than a colony with an older queen under the same conditions. Finally, remember that colonies going into winter which do not have access to stored pollen or are not able to forage for pollen will need to be supplied with a protein supplement.

Summer/autumn checklist

Summer maintenance starts from the end of April and continues through to the end of August, ending with the removal of the honey crop. The objective at this time of year is to encourage the colony to increase in size, without swarming, and then to build up stores of honey. This is also the time when the beekeeper will create any new colonies and add new queens for the next season.

MAY

- ☑ The flowers in May bring an increase in brood rearing. Make sure that queens are not left short of brood-rearing space. If necessary, rearrange your brood chambers by inserting empty, drawn comb into the brood area.
- ☑ Remove your chemical strips for varroa mite.
- ☑ Add a queen excluder and place honey supers on top of the brood chamber.
- ☑ Be on a close guard for swarming and up your inspections to once a week.

JUNE

- ☑ The early part of June marks the transition between the spring flowers and the blooming of summer flowers. Depending on where you live, nectar may still be fairly scarce and this is what beekeepers refer to as the 'June Gap'. Feeding may be necessary, especially if you have already removed the spring honey.
- ☑ If June is particularly dry, make sure that the bees have plenty of water available to them. A shallow dish near the entrance to the hive will suffice. They need water to mobilize their honey stores and to raise their brood.
- ☑ June is the month for swarming so keep a close eye on your hives and take the necessary precautions.
- ☑ Queen rearing should be undertaken at this time during a nectar flow. Any queens raised this month will get a good chance to mate during the warm days of July.
- ☑ If you are lucky enough to live near a supply of heather, you will need vigorous young queens

Below: *It is always a good idea to provide bees with additional water, especially in hot and dry weather.*

Above: *Siting beehives near heather will produce distinctive 'heather honey' with a dark colour and almost tangy flavour.*

and abundant foragers. Heather honey has been called 'the Rolls-Royce of honey' and it is something that you will either love or hate.

- ☑ Supers should now be ready for the main flow from white clover.

JULY

- ☑ By the third week of July the main floral honey flow from bramble and rosebay willow herb, will be drawing to a close.
- ☑ This is the time to watch out for robbing. Colonies who are desperate for stores will pounce on any bits of honeycomb from your inspections. Try waiting until late afternoon or evening before carrying out any manipulations.
- ☑ Continue regular weekly inspections to ensure the health of your colony.
- ☑ Add more supers if needed and keep your fingers crossed in anticipation of a great honey harvest.
- ☑ The swarming impulse will gradually die away towards the middle of this month. Should a late swarm occur, your colony will need assistance to establish itself adequately for winter.
- ☑ This is the month to start extracting your honey, and it may well continue through the next two months.
- ☑ Once the supers are removed, examine the colonies and if necessary treat for varroa mite.
- ☑ This is a good month to requeen.
- ☑ If you are rearing bees for heather honey, now is the time to start preparing for moving the hives nearer to the source, i.e. moors.

AUGUST

- ☑ Continue to watch out for robbing (see page 99–100). This occurs when the flow of nectar is greatly reduced. Be particularly careful when the hive is open and also remove supers from the apiary as soon as they are taken off the hive for processing.
- ☑ Remember – when removing the supers always leave one full one for the bees. This is to give them a good start on their winter storage.
- ☑ If you are raising a colony for heather, then now is the time to move your hives. Heather can make heavy demands on even the strongest stocks if there is prolonged poor weather. Watch their food supplies very carefully.
- ☑ Colonies will now be raising their winter bees and drones will be herded to the edge of the brood area and eventually expelled from the hive.
- ☑ Whenever the brood is inspected, look out for any abnormal signs and take appropriate steps to treat diseases. Treatment for varroa could be

paramount during August, as badly infested colonies can be swamped by hungry young mites.

☑ Any wet supers – i.e. ones that have some honey left on them after the bulk has been extracted – should be returned to the hive from which they were taken, to be cleaned up by the bees.

Below top: *Once they have fulfilled their purpose (fertilizing the queen), drones are driven out of the hive.*

Below bottom: *A mouse guard, placed at the hive entrance, allows bees in but keeps out mice.*

SEPTEMBER

☑ Colonies from the heather moors should all be returned home by the middle of the month. Examine the colony carefully and treat for varroa mite.

☑ Be alert for any bees from colonies that have collapsed because of an abundance of varroa mite. These bees could easily carry the mites into your healthy stocks. If these go unnoticed they will undo all your hard work.

☑ Check for other diseases. Any that you are not sure about, ask for help from your local beekeepers' association.

☑ All necessary feeding, uniting of stocks and general winter preparations should be completed by the end of September. Remove any remaining supers and queen excluders from the colonies.

☑ If you have fitted varroa floors, these should be carefully checked.

OCTOBER

☑ Brood rearing ceases during this month.

☑ The bees become less active and start to gather in their dense winter clusters on colder days.

☑ Stop liquid feeding after the middle of October and carry out final winter preparations.

☑ Reduce entrance size and fit mouse guards if necessary. Make sure your hives are weatherproof.

☑ Clean, sterilize and dry all extracting equipment. Repair any damaged equipment and treat stored combs against wax moth.

Surviving the winter

Now that you have completed the harvesting of your honey and your colonies have settled down into their winter state, you should have two main priorities. The first thing is to store your honeycomb and the second is preparing your hives for winter. First and foremost you will need to check your colony to make sure that it is free of disease, that it has sufficient bees to survive the winter and that there is a healthy, laying queen.

WINTER SITING

Some beekeepers like to move their hives to a different, perhaps more sheltered site, for the winter. You have to remember that moving a hive can be quite stressful to the bees, so once your hive has been relocated you will need to make a thorough inspection.

Try to choose a spot that will get as much winter sun as possible. The site should not be prone to flooding and the hive should be protected from prevailing winds and any extreme frosts. The sun is still vital to bees in the winter, as it is beneficial to the colony if they are able to fly from time to time before regrouping in a cluster.

HOW TO SURVIVE THE WINTER

Basically you will prepare your hives for winter on two separate brood boxes. The queen will shortly stop laying and the colony will go into their winter cluster for warmth and to maintain brood temperature. There will be no foraging trips, not only because of the cold but also because there will be no nectar sources. You may still see some activity, however, when the bees come out on a sunny day to defecate.

These are the key checkpoints to make sure that your hive is strong enough to survive the coming months:

☑ Has it got a laying queen?
☑ Are there enough bees – at least 15 frames for cold winters?
☑ Is the colony free of disease?
☑ Has the colony been treated for varroa?
☑ Are there sufficient stores to take it through the winter? If not, feed the colony.
☑ Has the queen excluder been removed?
☑ Are there two brood boxes?

Now you should carry out general maintenance on your hives to make sure they are in good condition to face the winter. Firstly, clean the floors,

Above: *Using a stethoscope helps to detect activity in the hive in winter without disturbing the bees and allowing them to become dangerously cold.*

scraping them or replacing with new ones where needed. Then check to see if the lid is watertight and that it won't be blown off in strong winds. If necessary, weight it down or strap it onto the hive body. Check that all the woodwork is sound and that it has no holes or splits that could let in the rain or any invading wasps. The bees will cluster in the empty brood area, so make sure that these frames are surrounded by frames of both pollen and honey. Make sure these are in the bottom brood box. You will find that the bees tend to move up during the winter, so the upper brood box should also have frames of stores. Any brood frames that contain brood can be placed in the upper box.

Towards the end of the autumn, the drones are starved and forced to leave the hive by the workers. They cling to the drones' legs and wings and generally harass them until they fly away or drop from the entrance. The drones are unable to forage for themselves and quickly die. If any colonies retain their drones, you will need to make a closer inspection as queen replacement may be incomplete. If this is the case, then the queen cells should be left intact and the bees left to complete the process. It is unusual for colonies who are tending queen cells to swarm.

THE CLUSTER

Because bees are cold-blooded, they have to find a way of surviving the drop in temperature during winter months. They do this by forming

Above: *A queen bee surrounded by workers. In cold weather they will huddle even closer.*

a cluster, or a 'huddle' of bees. Because they do not hibernate and have to continue feeding during the winter, they totally rely on the stores they have built up over the warmer months. As soon as the temperature starts to drop, the bees congregate and form a cluster that is dense enough to maintain a temperature of 32–34°C at the centre. The bees on the outer edges are tightly compressed, insulating the bees within the cluster. Because their body temperature is so much lower than the bees in the centre of the cluster, they need to constantly change position with the bees in the middle. By doing this the bees survive and form an intricate pattern of food sharing to keep up their strength.

If there is a warm spell, the bees will take advantage of this to move the cluster to another part of the hive containing honey stores. However, if there is a prolonged cold spell this can stop the cluster from moving and, having exhausted the supplies, the colony can die quite quickly even though they are just inches from fresh supplies.

The queen will remain within the cluster and move with it as it changes position. If the colony has sufficient supplies, they will start to feed the queen and stimulate her laying instincts towards the end of December or early January. Their aim is to replace any bees that have not survived the winter. The population will naturally decrease during the winter as the older bees die of natural causes.

The role the human can play in assisting in the bees' survival is to make sure that no moisture forms within the hive. Moisture that forms condensation can drip on the bees, and this can have a really disastrous effect on their natural ability to maintain the temperature. Make sure you provide adequate ventilation and ensure the hive is as watertight as possible. The ventilation shaft positioned just below the overhang of the roof should be left open during the winter and make sure the hive is positioned away from any strong winds.

SIZE OF WINTER COLONY

Most mature colonies will enter the winter with 40,000 to 60,000 bees. The population may dwindle to 10,000 to 15,000 bees by early spring as the older bees die. Brood rearing commences soon after the winter solstice, determined by the number of daylight hours. Generally speaking, a colony will not start to increase until late March or April. Young queens usually lay later and start sooner, but the supply of pollen during the autumn is a determining factor to the survival of the colony during the winter. It is advisable to unite smaller colonies to make one large one, as the best protection to bees during the winter is more bees. The general rule is to unite a colony which occupies six or fewer combs by the middle of September, first making sure that the weakness is not due to disease.

An easy way to unite colonies is by placing one brood box on top of the other with a sheet of newspaper in between. Alternatively, you can transfer weak colonies to nucs, or the empty frames can be removed and a dummy board used. It is possible to place two nucs under one single roof, as long as you make sure that their entrances face in different directions. Similarly, a weak colony can be placed over the crown board of a strong colony.

PROTECTION FROM ENEMIES

By taking the necessary precautions, almost all winter stock losses are avoidable, with the three main threats being starvation, the elements and bee enemies.

Robbing

Robbing bees, wasps and bumblebees can be a threat to a honey bee colony that is trying to survive the cold winter months. This usually takes place at the end of the nectar flow and weak colonies are particularly prone to robbing. Care should be exercised when feeding sugar syrup, trying not to drop any on the floor around the hive. It is also a good idea to reduce the entrances.

Robbing is often a problem when food supplies are short and stronger colonies rob the weak ones. Feeding seems to make this worse and can sometimes even set it off. When you see this start to happen you should reduce the entrances on all your hives as this will slow them all down. You need to remember to keep an eye on the situation and re-open the entrances when the flow returns.

It is quite common for queenless hives to be robbed more frequently than those with a strong, laying queen in place. However, you do need to make sure that it is actually robbing taking place and not just a regular group of your own young bees orienting on a warm, sunny day. They will hover and fly around the hive, just as the robbers do, but with time you will learn to recognize the difference. Young bees are fuzzy in appearance and are much calmer compared to the robbers. Look at the entrance. Robbers will be in a frenzy, whereas your own bees might be

Below: *You can deter robber bees with strong aromas such as Olbas oil and VapoRub™.*

busy but their movements will be organized. Wrestling at the entrance is also a giveaway sign, but be careful – lack of fighting at the entrance does not mean your hive is not being robbed.

One certain way of making sure that your hive is being robbed is to wait until dark and close the entrance. Any bees that show up in the morning and try to get into the hive are probably robbers, especially if there is quite a large number. If you know for sure that robbing is taking place, there are ways to stop it. The first way is to close the hive with some coarse hardware cloth for a day or two. This is only advisable if you know the colony is particularly weak. The robbers won't be able to penetrate the cloth and will eventually get tired of trying. It will help if you can feed your own bees and make sure they have a few drops of water as well during this period. After you open the entrance, make sure you reduce the size of the hole. You can also force the robbers to join the hive by closing up the hive, making sure that you feed, water and ventilate the colony for at least 72 hours. Another way is to fill the entrance with dried grass. The bees will eventually remove this themselves, but hopefully by that time the robbers will have given up.

Another useful tip is to rub some menthol vapour rub (for example, Vicks™) or a few drops of Olbas oil around the entrance, which will confuse the robbers because they can't smell the hive. It will not, however, confuse the bees that already live there.

Occasionally a weak hive will get totally robbed so there is not a single drop of honey left. The bees will quickly starve, so if you find you can't control the robbing it is better to combine some of your weaker hives and allow them to be robbed.

Mice

Mice can be a problem as bees are more lethargic once they have formed a cluster and will not fight off a predator. They build a nest in hive boxes, destroy comb in the frames and also eat holes in your equipment. In the autumn the beekeeper can keep mice out of the hive by placing a mouse guard at the entrance. They are sold by bee suppliers, but you could also make your own. If you can restrict the entrance to 6 mm, bees will be able to come and go but mice will not be able to enter.

Woodpeckers

Woodpeckers can cause considerable damage to a wooden hive. Netting or some kind of plastic skirting around the hive may help to deter them.

Humans

Even beekeeping is not safe from vandalism and theft, and it is becoming an ever increasing problem. Try to choose sites where there is no access to the general public and where the hives are out of sight of public footpaths and roads. Even though you will not be inspecting during the winter months, it is still advisable to make frequent visits to check that your hives are OK.

Above: *Mice can get into very small spaces, so use a mouse guard to deter them from wreaking havoc in your hives.*

Ants

Ants are a nuisance in the beehive. They often build nests under the top cover and above the inner cover where the bees don't bother them. They seem to cause very little damage to the bees themselves, and are probably more of a worry to the beekeeper than his colony.

Any chemical used to destroy ants would also kill your bees. If you have a major problem with ants, then you could stand the legs of your hive in cans filled with oil, which will prevent the ants from climbing up the side of the hive. Don't spend too much time worrying about them, though.

Wax moth

Wax moth can be a major problem when storing honeycombs but they can also be present in the hive itself. This little moth can often be seen running for cover when the roof of the hive is first moved. In recent years, the Greater Wax Moth (*Galleria mellonella*) has appeared in the southern counties of the UK. This is much larger than the Lesser Wax Moth (*Achroia grisella*) and consequently does more damage. The size of its larva is 2.2 cm compared to the 1 cm larva of the Lesser. It forms strong, silk-lined tunnels as it burrows through combs, feeding on discarded bee larva skins and cocoons. In addition to damaging your combs, it can also carve an unsightly boat-shaped cavity on the woodwork inside your hive, where it will lay its own young. It will occasionally chew the cappings from sealed worker brood exposing the bee pupa. This phenomenon is known as 'baldbrood' and can be a clear indication of the presence of wax moth in problem numbers.

Good housekeeping is the cheapest and simplest way of keeping on top of this pest.

- Make sure the floor to your hive is kept clean, especially the corners and joins.
- Keep combs and equipment free of brace and burr comb.

Below: *Wax moths are highly destructive to the fabric of the hive, the comb and the brood.*

- Keep the apiary free of discarded combs.
- Store old combs and cappings only briefly before melting down.
- Keep stored boxes of combs such as supers in a light place with good ventilation. Old queen excluders are ideal as a base for the stack. Low temperatures and good ventilation should inhibit the development of wax moth larvae.

Frost will kill all stages of the life cycle including wax moth eggs. Overnight storage in a freezer, if you have one available, will protect your combs for a while.

WINTER FEEDING

The reasons that bees need to be fed during certain times of the year are:

- To provide adequate stores for winter.
- To provide emergency stores in the season between colony inspections.
- As a means of administering drugs.
- To stimulate the queen to lay.
- To prevent colony starvation.
- To enhance wax production and the drawing of foundation and comb.
- When raising new queens and making up nuclei.

Bees dilute honey with water before they consume it. However, when nectar is stored, the moisture is removed. For these reasons sugar syrup that is fed for winter storage should be more concentrated than that which is used for spring feed.

Never use unrefined or brown sugar when making sugar syrup as this causes dysentery in bees. Sugar syrup has no smell to the bees and to make it more attractive you can add a little honey to give it an aroma. Never put an open container of syrup inside the hive or you will lose hundreds of bees from drowning. Most beekeepers use purpose-made plastic containers, which hold about 1 litre of syrup. It is best to place feeders in the evening when the bees have, or will soon, stop flying. This allows the initial excitement of the bees to subside overnight and reduces the risk of robbing.

To survive the winter a bee colony will need about 18 kg of food. One deep frame full of honey contains approximately 2.2 kg, and a shallow one 1.3 kg. Hefting the hive on a regular basis should give you a rough estimation of how much food is left. The best food is honey, but make sure you do not feed 'unripened' stores that have fermented as it may cause dysentery and aggravate nosema disease. If a super of food is left on the hive, it is recommended that the queen excluder is removed to avoid what is referred to as 'isolation starvation'. A good compromise is to supplement their winter honey with a feed of sugar syrup, which should be given in plenty of time for the bees to ripen and seal.

Preparing syrup for feeding

Generally there are two types of feed, a thick syrup for autumn feeding which will be stored more or less immediately and thin syrup for spring or stimulative feeding which is to be consumed without storing. The following are the recommended consistencies:

- Thick (for autumn feeding) – 1 kg sugar to 500 ml of water.
- Thin (for spring stimulation or pollination feeding) – 1 kg sugar to 1 litre of water.

Below: *Sugar syrup supplements the bees' diet in winter when other foods are less available.*

Honey bees winter well on stores derived from sugar syrup as these stores will not granulate. If the syrup is fed early enough – before the end of September in the south of England, for example – the bees will pack the stores around the brood nest where the bees will cluster. Remove the excluder and supers to alleviate the risk of them becoming contaminated with the sugar syrup.

Invert sugar

When bees collect nectar, they invert the sugars in the nectar by adding enzymes. The advantage of feeding invert sugar to bees is that it breaks down the sucrose into glucose and fructose, which are the two main components of honey. Since it closely resembles honey by being made up of the same two components, it is more easily digested by the bees and larvae and can be more readily consumed. Unlike regular sugar syrup, invert syrup is less likely to crystallize as it retains moisture for longer when used in pollen patties. It also helps to prevent mould and robbing is less prevalent when invert sugar is used for feeding.

Invert sugar is easy to make by simply adding some cream of tartar, lemon juice or dried active baker's yeast to the sugar syrup.

Below is a recipe for invert sugar. When making this invert sugar make sure that you boil the mixture for at least 20 minutes to invert as much of the sucrose as possible. Be sure to stir the mixture frequently and do not allow it to sit as it will caramelize in the bottom of the pan. Invert sugar has a shelf life of about six months.

Recipe for invert sugar

Ingredients
3.5 litres water
3.5 kg white granulated sugar
1½ tsp of cream of tartar

Method

1 Put the water into a large pot and bring to the boil.
2 Slowly add the sugar to the water and stir until the sugar has completely dissolved.
3 Add the cream of tartar and stir into the mixture.

Pollen or pollen substitutes

Bees' chief source of protein is pollen, and without it their brood would not develop and the colony would die out. It is therefore an essential part of their diet, also providing important lipids, minerals and vitamins. You might find you never need to feed pollen substitute, but if there is a dearth you might need to boost their reserves.

As a beginner, you will need to check the pollen reserves frequently. If these are sufficient, the bees will eat the pollen from them or from the pellets brought to the hive by the worker bees. As long as there is a plentiful supply within their foraging area, bees tend to manage their supplies well, but there are times when you might need to feed pollen or pollen substitutes, for example, when rearing queens.

The easiest way of feeding pollen to your bees is to place it in a shallow dish and put it under the lid of the beehive. Alternatively, you can take out three or four frames from the super and place the dish of pollen on the queen excluder. By doing this the bees will come and go and take the pollen as they need it. They will not store the pollen as they do with sugar syrup, but only feed on it when required. You can also make a sort of putty mixture by mixing the pollen with sugar syrup. If you wish to do this, place the pollen mixture on the top bars of the brood nest, or smear it over the surface of the queen excluder. This should last the bees for up to three weeks.

It is not advisable to feed pollen outside the hive, because it can encourage pests; it can spoil if it gets wet; and you also risk losing it on a windy day.

Pollen patties are another easy method of feeding. They can be bought ready made or prepared at home. For the latter, add pollen substitute to a third of a bucket of sugar syrup and mix until it is fairly stiff. When the mixture resembles a kind of dough, form it into patties.

To feed the bees, place the patties in paper bags, leaving the upper sides open. The patties can be put inside the hive even in quite cold weather; the operation is so quick it will not disturb the cluster. In early spring, a healthy colony can be expected to eat as much as 450 g of patty in one week, so start early to give your colony a boost.

Feeding hard candy

Some beekeepers prefer to feed hard candy instead of sugar syrup, as it minimizes the risk of moisture building up inside the hive.

To make hard candy you will need:

Ingredients

5 kg white granulated sugar
0.5 kg honey
1 litre water
¼ teaspoon cream of tartar

Method

1 Heat the water and add the sugar and honey. Stir continuously until the sugar and honey have dissolved completely.
2 Heat the mixture to 114°C but do not continue to stir. Once it has reached the desired temperature, remove from the heat and add the cream of tartar.
3 Leave the mixture to cool to 51°C and stir briskly until the mixture becomes cloudy.
4 Pour the mixture into a rectangular cake box or specially prepared candy feeder box. Once the mixture becomes solid it can be wrapped in waxed paper and placed inside the hive.

Calculating feeding quantities

It is very important to know the amount of food that a colony requires during the season. The beekeeper will need to check the supplies at each inspection and work out whether they have enough to last them through to the next inspection. It is worthwhile making a note that a flying bee uses approximately 10 mg of honey for every hour of foraging, bearing in mind that they can be out for as long as five hours. If your colony has approximately 13,000 foragers – say one third of the hive's total population – then the colony will require at least 4.5 kg of liquid stores. If, after hefting, you know that your supplies are low, then you may need to provide them with extra food.

This is also true of winter supplies, as many bees are lost due to starvation. Start your colony inspections in August and estimate the amount of food they require. Each frame within the brood chamber should be inspected and the amount of liquid stores worked out on the basis that a frame, when full and sealed with honey, weighs approximately 2.2 kg.

A healthy full-sized colony of bees, with an adequate supply of honey and a strong queen, can survive a hard winter. With your work done, you can now sit back, secure in the knowledge that you have done everything possible for your bees to make it through the winter. You can also utilize your spare time by repairing and painting equipment.

Left: *Giving additional feed in winter helps the colony to survive.*

Winter checklist

NOVEMBER

☑ There will not be much activity this month and the cold weather will send the bees into a cluster.

☑ Make sure that the hives are secure against the weather, particularly after storms.

☑ Store your equipment away for the winter.

DECEMBER

☑ The bees will now be in a tight cluster, so no peeking.

☑ If sunshine follows snow, shade the entrance to the hives so that the bees are not lured out into the cold, where they will rapidly chill and die.

☑ Repair and paint equipment.

☑ Clean supers, hive bodies, covers and frames of burr comb and propolis.

☑ Cut all combs with more than six square inches of drone cells from the frames.

☑ There is nothing else you can do this month, so enjoy Christmas and look forward to spring.

JANUARY

☑ The colony should be left in peace during this month. The queen is surrounded by her workers who are in the middle of their winter cluster. There is little activity except on a warm day, when the workers will take the opportunity to make cleansing flights. There are no drones in the hive at this time, but some worker brood will begin to appear in the hive.

☑ The bees will consume about 11 kg of stored honey this month, so checking stores is very important. Wait for a fine day before checking, though. This is done by hefting the hive (lifting one side and feeling the weight). With experience the beekeeper will be able to tell if there are enough supplies to keep the bees going for another month. If the colony feels light it should be given a feed of candy. If the weather is abnormally warm, a frame with sealed honey can be added if there is one available from the hive last year.

☑ Check that the hives are still upright and that your mouse guards are still in place.

☑ Check for any damage from woodpeckers etc. and ensure that there is plenty of ventilation.

Above: *Use the early months of the year for essential maintenance of your hives.*

- ☑ If you have had quite a bit of snow and the ground in front of the hive is affected, clear the immediate area so that the brightness does not tempt the bees out into the cold air. However, if there is snow covering your hive, it is OK to leave this as it will act as an insulator.
- ☑ Clean, paint and repair equipment.
- ☑ Check the apiary for wind damage.
- ☑ Make a plan of the coming season's tasks.
- ☑ Read catalogues and order equipment for the new season.
- ☑ Order packaged bees (if required) from a reputable supplier.

FEBRUARY

- ☑ The queen, who is still nice and cosy in her winter cluster, will start to lay a few more eggs each day. It is still only females in the hive and workers will start making more cleansing flights on mild days. The bees will consume about 11 kg of honey this month.
- ☑ Check that there are sufficient stores by hefting.
- ☑ Towards the end of the month the floor should be replaced with a clean one and the debris checked for varroa mites.
- ☑ If there are a significant number of varroa mites (say over 100), the colony should be treated.
- ☑ Make sure that supers and other spare equipment are in good order and have been treated with preservative. This work needs to be done in the winter months, because from April onwards you will be very busy and in need of your extra equipment.

CHAPTER FOUR

HONEY AND OTHER HIVE PRODUCTS

The main reason we set up, maintain and stock a beehive is to harvest the honey. Honey is one of nature's wonders and it has been a treasured commodity for centuries. It is one of the purest foods found in the world, it is easily digested, it is a powerful source of energy and, quite simply, is delicious.

The golden harvest

Bees have been producing honey for their own food sources for over a million years. With a little encouragement, the honey bee will produce far more of this golden nectar than they need and man can take advantage of this glut.

Honey bees obtain two things from flowers – nectar for honey production and pollen. Nectar is the clear liquid that drops from the end of a flower blossom, which is basically 80 per cent water with some complex sugars. As the female worker bee sucks nectar from the flowers, it is stored in her special honey 'stomach'. This is completely separate from her digestive stomach, although there is a valve connecting the two which she can open when hungry.

When her nectar stomach is full, she will return to the hive with her load. A worker bee returning with nectar can easily be confused with other workers returning with pollen, but they never collect both on the same trip. Worker bees also have hair-like baskets on their hind legs, and it is these that are filled with pollen grains while they visit flowers. The bees make quite a comical sight as they waddle about the hive with their oversized loads of pollen. This pollen, sometimes referred to as 'bee bread', is mixed with honey and forms the diet of the next cycle of brood being raised. When you carry out your inspections, you should be able to see a multicoloured ring of pollen encircling the brood in the middle of the comb.

Below: *Her pollen sacs full to bursting, a worker bee collects more pollen from the flowers of Canadian goldenrod.*

When a bee returns with a full load of nectar, which is around 70 mg – almost as much as a bee weighs itself – it is met by other workers who are ready to relieve her of her load. This is done by mouth to mouth transfer between the field bee and the hive bee. They have extended tongues specifically for this purpose. The recipient bee then processes the honey in its mouth and honey stomach by adding enzymes that break down the complex sugars in the nectar into a more digestible form.

After the nectar has been processed, the bee deposits small droplets on the upper side of the cell wall where it will finally convert into the viscous honey that we are accustomed to. This conversion is largely due to evaporation which, in turn, is hastened by the warm temperature maintained in the hive. The end result is a thick, viscous honey with only a 17 to 18 per cent moisture content.

Above: *A forager bee transfers nectar to a house honey bee.*

WHAT GIVES HONEY ITS FLAVOUR?

The aroma, flavour and colour of honey depends entirely upon the type of flower from which the nectar was collected and varies dramatically. The majority of honey comes from nectar sourced from a variety of different flowers and these are known as 'polyfloral'. Some plants, however, manage to produce enough nectar during their short flowering season, to enable the bees to feed entirely on one type of blossom. This honey is known as 'monofloral' and is keenly sought by beekeepers and consumers alike, as it has a unique flavour and aroma.

Honey is produced all over the world, from the heat of the tropics to the cold districts of Scandinavia, Canada and Siberia. In Britain, honey is produced primarily for the local market with well-known flavours like apple blossom, cherry blossom and heather, but we are very reliant on the climate and the area in which the bees are kept.

HOW IS HONEY SOLD?

Honey is sold in four different forms:

1. Honeycomb or comb honey

Direct from the hive as honey-filled beeswax comb as it is naturally stored by the bees.

2. Liquid honey (extracted)

This is prepared by cutting off the wax cappings and putting the comb in a honey extractor.

3. Creamed honey (granulated)

This is made by blending one part granulated honey with nine parts liquid honey. The mixture is then stored at about 14°C until it becomes firm.

4. Chunk honey

Pieces of comb honey in a jar with liquid honey poured around it.

1.

2.

3.

4.

Extracting the honey

If this is your first year as a beekeeper, then you shouldn't expect too much from your honey yield. The colony is still trying to establish itself, but your patience will pay off as the second year will be much more rewarding. It is difficult to determine which flowers your bees will forage from, so expect a blend from many different nectars. Your honey can be classified as 'Wildflower' honey, should you decide to sell it.

Below: *1. A honey extractor; 2. Uncapping the wax cells; 3. Filtering extracted honey.*

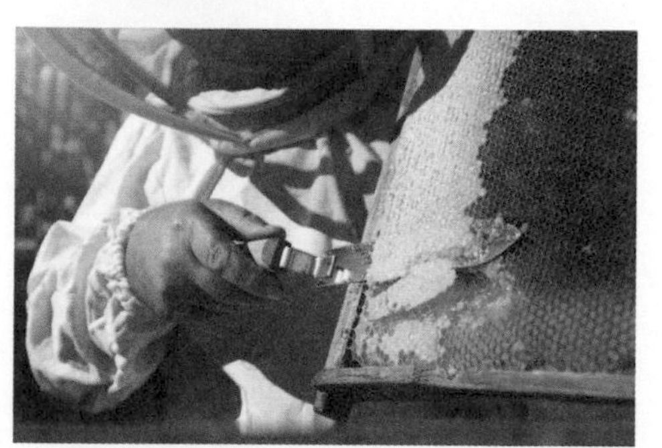

THE TIME TO HARVEST

The key to honey harvesting is to begin when the stores have been capped over by the bees, towards the end of the season. By this stage the honey will have been matured by the bees and sealed up, which means it is ready for us to eat. If at least three-quarters of the honey is sealed, it is time to harvest.Any earlier than that and the water content will be too high and it will ferment in storage.

You will need to estimate how much honey you think you will be extracting and you can assess this by checking the honey stores in the brood chamber. A deep frame full of honey will weigh approximately 2.7 kg and the bees will need ten of these to survive the winter. Two shallow frames will equal one deep frame. Now you need to make sure you have all the equipment ready (see pages 40–42):

- an extractor
- an uncapping knife or machine
- filters
- food-grade containers

DEALING WITH THE BEES

Now you have all your equipment ready and you have found a suitable room for carrying out the extraction, it is time to go to the hives and remove the supers. The only problem you have now is that there will be thousands of bees on the combs and somehow you need to get rid of them.

Assuming you have regularly been checking for American foulbrood and European foulbrood, you can now use one of the methods suggested below to deal with your bees.

Bee escape boards

Many beginners use a bee escape board which is placed between the brood boxes and the supers. The bees go down to the brood box at night through the escape board, but are unable to pass through it again to reach the supers. They are very basic in design, with a hole or several holes in them. Various devices can be placed over these holes to only allow one-way access to the bees. These only need to be put in place overnight and removed again once you have taken out the supers.

There are many different designs to choose from, but whichever one you go for, make sure it has several escapes so that if one becomes clogged up, the bees have an alternative route. If it has been a particularly warm night you might find there are still quite a few bees left on the combs. The other disadvantage to using this method is that you need to make two visits to the hive, once to place the board and once to remove the supers, causing more stress to your colony.

Fume boards

A fume board is exactly the same size as the lid of the hive, except that it is covered with tin. Within these boards is an absorbent fabric which is designed to soak up a liquid bee repellent. The tin is usually black in colour so that it absorbs the warmth from the sun. This warmth will help the repellent evaporate more quickly. You can buy the repellents from any reputable bee supplier. If you place the boards over the top of the frames in the top box, the bees should start to move down quite quickly, running away from the odour of the repellent. Puffing a little smoke in also helps to encourage them to start moving before you place the board.

It is important that you always follow the directions carefully, because if you use too much repellent the bees can become confused and start to cling to the comb. This makes the beekeeper's job even harder. After a few minutes the majority of bees should have moved down. Then you can remove the super and place the fume board on the next board until you have

Below: *Placing a fume board on top of the hive will encourage the bees to move down inside the hive.*

Above: *A soft-bristled bee brush can be used to remove bees from the comb but you need to be careful not to irritate the bees when brushing.*

worked your way through the hive. Remember if you have made any extra holes in your honey super boxes, you will need to block these up so that the fumes do not escape.

Bee brushes

Many people who only have a small number of hives use bee brushes to gently brush the bees off the combs. Bee brushes can be bought relatively cheaply from bee suppliers, and are specially designed for this purpose. This method has two main drawbacks. One, the brush can quickly become clogged with honey and two, the bees often become quite defensive. By far the easiest method is to have an empty super available. Remove the frame from the hive, shake it to remove as many bees as possible and brush the rest off in front of the entrance to the hive. Place the honey frame – now free of bees – into the spare super and, once full, cover it up.

Motorized bee blowers

This is a motorized device, similar to leaf blowers, using forced air to blow bees from supers when harvesting honey. They are often used by commercial beekeepers, but you probably won't want to go to the expense of this piece of equipment if you are starting out. The blower is used in conjunction with a stand, on which the supers are placed. The stand is placed on the ground in front of the hive and the bees are blown downwards out of the super so that they can crawl back into the hive.

Transporting the supers

Once you have made sure that your hives are disease-free and you have successfully removed the frames from the supers, now is the time to transport them to the place where you are going to carry out the extraction. If this is quite a distance from your apiary, then you might consider moving them by truck or car. Bear in mind these are quite heavy and you might need some assistance, especially as you will probably be surrounded by a lot of bees who want to get their honey back.

When supers are removed from the hive, some of the comb will break and start to drip and leak. You might consider putting the super inside a large plastic garden refuse bag, which will not only contain the honey but stop the bees from robbing it as well.

Whatever method you choose to transport your supers, make sure they are secure and that you place them in a safe place away from prying bees before you start your extraction process.

The extraction process

You now have your boxes full of honey frames safely in your kitchen, garage or shed, carefully sealed off from bees trying to retrieve their precious nectar. Beginners tend to choose the kitchen for their first extraction, as it is usually already clean and sterile. Now is the time to start the final process of extraction.

Have your extractor ready and waiting and a kitchen knife in a bowl of hot water to heat up the blade. The boxes should be placed on boxes on the floor, which should be protected by newspapers or plastic sheeting, as the combs are bound to drip. The final thing to do at this stage is to place an empty super next to the full one and you can get started.

Extracting

Take the first frame out of the box and hold it over a large bowl or container. This bowl needs to be large enough to catch the cappings without them dropping on the floor and ideally a strainer is in place within it to allow some of the honey to drain from the cappings.

Above: *Scraping off the wax cappings to allow the honey to be collected from the comb.*

It is a good idea to have a bridge across the top of the bowl, into which a nail is driven in the centre. The frame can then be balanced on the nail while you are decapping, and turned easily to access the other side.

Run the hot knife over the surface of the comb so that it slices off the wax cappings that are covering the honey cells. Cut the cappings as close as possible to the surface, leaning the comb to one side to allow them to fall away from the frame.

Now turn the frame over and do exactly the same thing on the other side. As you finish each frame, place it in the compartments of the extractor until you have removed the caps from all the frames.

Always make sure that you load the extractors evenly as uneven loads could cause the extractor to move about across the floor.

If you are considering selling a small amount of your honey, you must make sure you use a stainless steel or food grade polythene extractor. There are many old tin extractors on the market, but it is advisable to leave these well alone. Borrow or hire a good extractor, until you feel you can afford one yourself. This will also give you the experience of which sort of extractor you would like to buy.

Now the extractor is full, you need to use a bit of elbow grease. Turn the handle slowly at first and build up speed as the frames empty of honey. Continue in this way until the job is finished. If you are using a tangential extractor, you will need to turn the frames round and spin again.

Spring or summer flow

If you are extracting the spring flow, replace the supers on the hives for the bees to clean and refill. If you are extracting the summer flow, then the supers must be stored for the next year.

There are two things that you can do with these sticky supers. Either replace them on the hives for the bees to clean up, then store them in a shed, outside, or on the hives over the crown board making sure that mice cannot get access to them. You can also store them 'wet' in a bee- and mouse-proof place. It is said that putting wet supers on the hives in spring encourages the bees to enter them. On the downside, though, it is possible that the honey left in them could ferment and smell nasty.

THE HONEY

The honey in the base of the extractor will stay there until the tap is opened. Once the level of the honey reaches the level of the spinner, you will need to put the honey into the honey buckets. Honey should, if possible, be strained directly from the extractor but if it has started to granulate in the comb, it will not go through a fine strainer. If this happens, it can either be warmed immediately, strained and stored in buckets, or it can be run straight into buckets. In this case, before it is bottled it must be warmed until it becomes liquid, and then strained.

If your honey is for your own consumption you will not need to be too fussy at this stage, as a clean-looking product is perfectly satisfactory even if it does still contain a few particles. Honey that is bottled immediately after extraction can set very hard in the jar and be difficult to remove. To avoid this, first store it in honey buckets and, when it is required, warm it gently until it is runny enough to bottle.

It is quite easy to build a warming box for your honey. Make a wooden box large enough to hold

Below: *Extracted honey contains all kinds of debris. Filtering it will improve the clarity and taste of the final product.*

the honey bucket, and heat the box using two 40 watt bulbs. To warm crystallized honey so you can bottle it, place the bucket in the warming cabinet at 32°C for three to five days. When stirred this will be a good consistency for bottling. You can also make honey runny again by placing it in a water bath in a cool oven. Loosen the lids of the jars slightly first. You can also liquefy the honey by using a microwave on the lowest setting, but make sure you do not overheat as this will spoil the flavour.

If you want to reduce the honey to a clear liquid, the temperature will need to be 52°C for two days. You might find you have to experiment, as larger containers will take longer than smaller ones. Be careful not to overheat the honey, though, otherwise it will end up tasting like toffee!

After you have finished extracting, the mush of the cappings and honey can be put into the strainer to allow as much honey as possible to drain from them. The washing water from the cappings when they are turned out from the straining cloth, forms a good basis for a brew of mead if you are into your home brewing.

When you are completely finished with your extractor, wash all the equipment, starting with cold water so that scraps of wax float away and do not melt and stick to everything.

Testing your honey for purity

Make sure that once your honey is in the buckets or containers, that they are tightly capped. Honey is hygroscopic, which means that it will attract moisture from the atmosphere. It is a good idea to test the moisture content of your honey, as it can easily ferment. This could result in the honey exploding in the jar which will result in a foul taste. You can either go to the expense of buying a honey hydrometer, or borrow one from a fellow beekeeper. You will need to make sure that the hydrometer is correctly calibrated, otherwise you risk getting untrue readings.

Alternatively you could carry out the following simple tests to ascertain the purity of your honey:

You will need:

- One or more samples of honey
- Some cotton wick
- A candle
- Some matches
- Glass of water
- Blotting paper

1 Pour half a cup of honey slowly into another cup. Pure honey will spin clockwise as you pour it. This is a result of the asymmetrical structure of honey molecules that gives it a right-hand bias.

2 Test the water content by dipping a length of cotton wick into the honey. Hold one end to allow excess honey to slowly dribble off. Light the candle and then hold it to the soaked end of the cotton wick. If your honey is pure, it will contain minimal water content and consequently the wick will burn. If it is impure, it is likely to have a high water content which will prevent the wick from burning.
3 Try to dissolve a tablespoon of honey in water. Pure honey will not dissolve, but rather form a lump and sink to the bottom of the glass. Impure honey will easily dissolve in the water. You can also use methylated spirit instead of water. Impure honey will dissolve and turn the methylated spirit a milky colour.
4 Pour a teaspoon of honey onto a piece of blotting paper. Unlike impure honey, pure honey will not be absorbed by the blotting paper. A coffee filter works well for this purpose, too.

Hydroxy-methyl-furfuraldehyde (HMF)

If you are intending to sell your honey, then you will also need to test it for HMF, because only a certain percentage of this is allowed. HMF is a breakdown product of fructose (one of the main sugars in honey), which is formed slowly during storage and very quickly when honey is heated. The amount of HMF present in honey is therefore used as a guide to storage length and the amount of heating which has taken place. The amount of HMF present will depend on the variety of honey. High levels could indicate excessive heating during the extraction process. Some countries set an HMF limit for imported honey; in the European Union, for example, a level of HMF above 40 mg per kilogram is not legal.

Below: *The kind of flowers from which the bees have drawn nectar will affect the finished colour of the honey.*

Colour

Honey is colour graded into light, amber and dark categories, which do not have any bearing on the quality. Honey colour is measured on the Pfund scale in millimetres. While it is not an indicator of the quality of the honey, there are exceptions to the rule. The darker the colour, the higher its mineral contents, the pH readings and the levels of aroma and flavour. Minerals such as potassium, chlorine, sulphur, iron, manganese, magnesium and sodium have been found to be considerably higher in the darker honeys.

SECTIONED HONEY

One of the simplest methods of cropping honey is in specially prepared sections a little over 10 cm square. It is eaten with the wax comb, and is one of the best ways to present honey, as the aromas and flavours are unimpaired by extracting and heating. To use the sections you will need to buy a section crate to put inside the hive or to take the place of a super. The sections are then fitted inside it with each row of three separated from its neighbour by a divider. The sections themselves usually come flat, which need to be folded up and fitted with starters of thin wax foundation. To prevent the wood from breaking up when you fold it, make the outsides of the folding joints damp first.

Make sure you insert the foundation the right way with pairs of opposite parallel sides of the hexagonal cell bases lying vertically and not horizontally, and the points of the hexagons at the top and bottom, not at the sides. The sections are fitted inside the crate and when it comes to harvest the honey, they are simply removed from the box. After removing any surplus propolis, the sections are now ready for sale. If you wrap the sections neatly in cling film, it not only keeps them clean but enhances their appearance. The main disadvantage

Right: *Chunks of honeycomb are a great way for the beginner beekeeper to get into honey products.*

to this method is that you lose the value of the beeswax cappings (see page 127), as the honey is sold sealed in their individual frames.

CUT HONEYCOMB

Another method of making comb honey is to use full frames of non-wired, thin foundation and, when you come to harvest the honey you cut the comb using a comb cutter. These cutters are relatively inexpensive to buy and look just like large biscuit cutters.

The comb is usually packaged in specially designed boxes and is eaten with the wax combs. This is probably one of the best ways to present honey, as the aromas and flavour are not impaired by extracting and heating. This method is attractive to the beginner who does not have access to an extractor, as very little equipment is needed and it doesn't take up a lot of time.

MIXING COMB WITH HONEY

This is a good way of using up odd chunks of honeycomb. You simply place the chunks in an empty jar and then fill with honey.

CREAMED HONEY

Creaming honey is simply controlling the natural crystallization process. Nearly all honeys will granulate if left, and in the UK it is referred to as 'set honey'.

The speed and texture at which the honey granulates is mostly a product of the ratio between the two main sugars in honey – dextrose and levulose (also known as glucose and fructose). If a honey has a high dextrose to levulose ratio, it will granulate quickly with a fine crystal. The other way round and it will take longer, with crystals that can feel sharp against your tongue. To 'cream' the honey, you will need to mix in a percentage of honey that you already know has granulated finely. This is what is referred to as a 'starter'. To speed up the granulation, the starter needs to be mixed thoroughly with the liquid honey and then the container must be kept cool – not in a refrigerator, but ideally at around 14°C. By maintaining this temperature, the honey will granulate quickly and retain the same consistency as the starter. It will need to be stirred occasionally during the process. Once it turns cloudy, it can be put into its final containers. It will still need to be kept cool to assist the final granulation.

Manuka honey – a little miracle

Manuka honey is a special type of monofloral honey which is produced by bees that gather nectar from flowers that grow on the Manuka bush (*Leptospermum scoparium*). It is a wild shrub which is indigenous to New Zealand, growing on the North and South islands. Studies have proved that Manuka honey contains very powerful antibacterial, antimicrobial, antiviral, antioxidant, antiseptic, anti-inflammatory and antifungal properties, making it extremely effective in the treatment of a wide variety of health conditions.

Although regular honey has many healing properties, it loses most of these once it comes into contact with wound fluids, or if it is exposed to light or heat. Manuka honey, on the other hand, has additional antibacterial properties that have the ability to destroy the bacteria that are responsible for infecting wounds and can even work into deep skin tissue. All types of honey contain hydrogen peroxide which is produced when bees add enzymes to the honey. In regular honey, hydrogen peroxide is produced in a slow release way that is strong enough to be effective in destroying bacteria but not enough to harm tissue. Manuka honey differs in that it also contains plant-derived components such as methylglyoxal and what some experts refer to as the Unique Manuka Factor (or UMF).

MEDICINAL USES

Based on recent studies, Manuka honey has been known to effectively treat:

aches and pains • acid reflux disease • acne • arthritis • athlete's foot • blisters • burns • chronic wounds • cold sores • eczema and dermatitis • gum disease • infections • insect bites • MRSA/staph infections • nail fungus • pressure sores • psoriasis • rashes • ring worm • skin ulcers • sore throat • stomach ulcers • surgical wounds • wounds and abrasions • wrinkles

Added to this, Manuka honey is effective in destroying a range of bacterial and fungal infections – a little miracle indeed.

Wax, propolis, royal jelly and pollen

In addition to collecting nectar and pollen from flowers, bees also gather the sticky sap from poplar and evergreen trees. Bees process this sap into a substance called propolis, which they use, along with the wax, to make, defend and repair their hives.

Although honey is the main commercial product that humans harvest from beehives, other bee products such as wax, propolis, pollen and royal jelly are also collected for human use. Before deciding whether you want to harvest these other products, you will need to find out about the compliance issues which apply to these items, in the same way as honey itself. These compliance rules are becoming tougher each year, and are governed by the main consumer blocks in the EU and the USA. Of course if you are only going to use them for your own purposes, then you need not worry.

BEESWAX

Beeswax is a secondary by-product of extracting honey, but still valuable to beekeepers. It can be used for candles, furniture polish, creams, cosmetics and a whole range of other uses. Additionally, it can be turned into sheets of foundation to replace old worn-out ones. Over a period of time, the comb darkens and eventually the beekeeper will decide to add a new frame and recycle the old wax.

Recycling can be done using several different methods. The simplest, and one which many beekeepers use, is a solar wax extractor. This is usually a wooden box, with a glass lid, very similar to garden frames used for propagating new plants. The frames containing the wax are placed inside the box and, as the sun heats the inside,

Below: *Golden blocks of beeswax prepared ready for candle-making.*

the wax gradually melts and runs to the bottom where it is collected in a tray and then into a container. Once the container is full it can be removed. If you lay kitchen paper or a piece of fabric at the bottom of the tray, it acts as a filter. You will probably need to filter your wax several times to get rid of impurities such as propolis. The value of beeswax varies according to its purity and colour. Light-coloured wax is more highly valued than dark, which usually means that it has been overheated or contaminated.

Above: *Perhaps the simplest method of wax extraction – simply scraping off the cappings.*

Another method of extracting wax is to use a steamer, which will become a necessity if you decide you want to build up your hives to a commercial level. This is fairly self-explanatory, with a mesh basket at the top to hold the comb, water in the bottom and some form of heat underneath. A portable gas ring powered by a cylinder of butane is ideal if you are nowhere near a power supply. Alternatively an electric element works just as well if you have a mains supply in your apiary.

Perhaps the simplest way of all is to render down old combs by putting them in a clean sack and plunging them in a container of boiling water. As the sack boils away, you can press down on it with a block and the molten wax will escape and rise to the surface. A spout near the surface allows you to pour the molten wax into a container. Very often beekeepers will use a solar extractor to do the final filtering as the last two methods are not very proficient at this.

BEESWAX FROM CAPPINGS

The finest wax comes from the cappings, which are naturally pure and white. Depending on the method of uncapping, the amount of honey mixed with the wax cappings will vary. You can either leave these to drain in a filter or a sieve, or alternatively hang them up in a muslin bag to drain. Using a heated tray while uncapping, means the wax and honey can be separated and processed at the same time, cutting out a lot of the sticky work. The following is a simple method of rendering which can be carried out in your own kitchen if you don't have any specialized equipment in your early days as a beekeeper.

1 Put all cappings in a large stainless steel pot or saucepan. It is best to keep this pan specifically for this purpose as you will not want to cook in it again afterwards. Cover the wax with water and put the saucepan on your cooker with the heating set to maximum. You need it to boil rapidly, so make sure you cover the pan with a lid, so that pieces of hot wax don't jump out onto the top of your cooker. Keep a constant eye on it as you don't want it to boil so fast that it starts to boil over making a nasty mess. Boil continuously for about 30 minutes.
2 Now it is time to strain off the liquid, but remember your wax will remain in that liquid so it is advisable to use a wax screen at this stage. If you do not have one, then you can use several layers of muslin and make sure you seal it securely over the top of the pot you are about to transfer to.
3 As soon as you remove it from the stove, start to gently pour the liquid through the strainer into the new pot. It is quite a good idea to do this outside, as wax is very difficult to get off your kitchen floor if you spill it.
4 Once the liquid starts to cool, the wax will float to the surface and begin to harden. This process will take about four hours, so it is best to leave it overnight until it is nice and firm. Don't worry if you still see some foreign particles in the wax at this stage, as you still have a little more rendering to do.
5 Remove the wax from the top of the liquid and throw the liquid away. Break the wax up into small pieces and put it in a stainless steel pan to melt down again. A double boiler is ideal for doing this as you do not want to melt over direct heat or you will scorch the wax black and it will be ruined.
6 Once it has melted down completely, slowly pour it through a piece of muslin into a clean container. If the muslin starts to clog up, it is best to start with a new piece. Never try to squeeze the muslin as you will end up with nasty pieces of sediment in your wax and you will have to start the rendering process all over again. Now you should be left with a beautiful golden yellow wax which can be sold or used to make your favourite things.

Reusing beeswax

If you want to reuse your beeswax in your hive, it is worthwhile investing in a wax foundation mould. This is quite simply a press with hinges to make a sheet of wax with the natural honeycomb patterns of the bees on its surface. To make a sheet of foundation you simply melt the wax in a double boiler or bain-marie, spray some liquid lubricant over the rubber leaves inside the mould, pour over the wax and close the press. Wait for a

Left: *The sweet, natural scent of beeswax candles makes them ideal to burn when you want candlelight at a dinner table but don't want to compromise the taste of your food.*

while for the wax to solidify, trim off any excess wax, open the frame and gently pull off the sheet of wax. Although this will be slightly thicker and a little more brittle than a commercially made sheet of foundation, it will be just as functional.

PROPOLIS

Sometimes called 'bee glue', this material has many properties that both bees and humans use. Bees use propolis for:

- A building material to decrease the size of nest entrances and to make the surface smooth for passing bee traffic.
- To keep the hives dry, draught-free, secure and hygienic.
- To seal up any cracks where microorganisms could flourish.
- To varnish inside brood cells before a queen lays eggs in them, providing a strong, waterproof and hygienic unit for developing larvae.
- To embalm the bodies of mice or other predators, which are too large for bees to eject from the nest. If they did not do this, the bodies would decay and could become a source of infection.
- As it contains volatile oils, propolis is also used as a kind of antiseptic air freshener inside the hive.
- The wild Asian honey bee, *Apis florea*, uses rings of propolis like bands of grease to coat the branch from which its single comb nest is suspended, as a protection from predators.

Humans have known of propolis since Stone Age times, when it was used as glue to secure flint arrowheads or spear points. In more recent times, it has been used as a component of the varnish on prestigious violins. Propolis has been recognized as a medicine for centuries as it has been proved scientifically to kill bacteria. It is also a common ingredient in toothpaste, soap and various ointments. Dissolving propolis in alcohol makes a tincture with many claimed medicinal properties.

Above: *Using a propolis grid or screen means that the bees fill in the propolis themselves.*

Harvesting propolis

There are two basic methods of harvesting propolis:

1 Scraping. You can scrape propolis off the woodwork, but the main problem with this method is that you usually get small scrapings of wood in the propolis. It is also very time-consuming, so it is probably advisable to invest in a propolis screen.

2 Propolis screens (also called propolis grids) can also be used to collect the raw product. These screens are similar to a perforated sheet of queen excluder, except that it is made of a polythene material 3 to 4 mm thick. Short round-ended slots approximately 4 mm in width are punched into the plastic and it is these slots that the bees fill with propolis if the screen is placed on the hive instead of a crown board. When these are full, you remove the screen and place it in a freezer. When it is frozen, remove the screen and flex it slightly so that the lozenge-like pellets drop out. The flexing action and the soft nature of the sheet material produce sloping sides to the slots which aid in the removal of the pellets.

You can also use polyethylene-yarn netting if you don't have a screen. The propolis is removed by scrunching up the netting after it has been frozen, in much the same way as described above. This netting can either be bought from your local bee supply shop or from a DIY or garden centre.

Remember to keep the propolis well sealed after collection, as it is prone to wax moth.

ROYAL JELLY

In monetary terms, royal jelly is the most valuable product of the hive. It is the food of the queen bee larvae and, by feeding a worker bee larva this substance, she will turn into a queen. Royal jelly is secreted from the glands in the heads of young worker bees, which has first been mixed with some sugars and proteins from the worker bee's stomach. It is a mixture of different components, including proteins, sugars, fats, minerals and vitamins and is a thick, white liquid not dissimilar to fresh yoghurt.

Above: *Scraping propolis from the top of the frames takes a while and gives a low yield.*

Producing royal jelly

Although the collection of royal jelly does require skill, any beekeeper who has reared their own queens will know how to produce it. Under natural conditions a larva destined to become a queen bee develops in a large wax cell or cup. Inside this cell the worker bees place large amounts of royal jelly. This is when the beekeeper starts to manipulate the hive to encourage the workers to produce a large number of queens – perhaps 50 or more. This stimulates the worker bees to produce vast amounts of royal jelly – you will need to feed your colonies extra sugar syrup to achieve this – and place it in the queen cells. However, instead of the larvae feeding on the royal jelly and developing into queen bees, the larvae are removed and the royal jelly is harvested by the beekeeper.

Below: *Propolis oil may be used for cold sores, and swelling and sores inside the mouth.*

Removing the royal jelly

The easiest way to remove the royal jelly from the cells is to use a small suction device which is available from bee-supply shops. Royal jelly must be harvested under hygienic conditions and rapidly refrigerated, frozen or freeze-dried.

Timing is the main criterion when harvesting royal jelly – leave

Above: *A queen bee cell containing royal jelly.*

it too late and the cells will be capped and the queen bee larva will start to develop and it won't be worthwhile extracting what is left. It is a skilled procedure, but one that is well worth mastering, so obtaining guidance on this matter will increase your chances of success.

POLLEN

Bee pollen is the male seed of a flower blossom which has been gathered by the bees. The bee mixes its own digestive enzymes with the pollen and it has been referred to as nature's most complete food. Humans have regarded pollen as a highly important health-food product for centuries, as it contains most of the known nutrients necessary for human survival. When compared to any other food, it contains a higher percentage of all necessary nutrients which is said to improve energy, stamina and strength and general wellbeing. For these reasons, bee pollen is extremely saleable and some beekeepers dedicate their hives to pollen rather than honey collection.

You should only consider collecting pollen from really strong, disease-free colonies when it is in natural abundance. Bees collect pollen avidly during the spring months when the colony is expanding rapidly. Depending on where you live, though, pollen could be in abundance at other times of the year.

Below: *Like propolis, royal jelly products are claimed to have a number of impressive health benefits.*

Trapping pollen

Special traps for collecting pollen from the legs of honey bees have been designed for this purpose. These traps vary greatly in size, appearance and method of installation on the hive. All traps, however, have two basic elements:

1 A grid through which pollen-carrying bees must crawl to separate the pollen pellets from their legs.
2 A container to store the pellets.

Above: *Bees collect pollen throughout the year but most avidly in springtime.*

A pollen trap should catch between 60 and 80 per cent of all the pollen brought to the hive. Traps should be designed to exclude all debris – insect parts, wax moths etc. – and should be easy to operate and fit. They protect the pollen from sunlight and moisture and should not restrict the ventilation into the hive. They should also be judged on how easy they are to remove and clean as they can quickly become clogged. The size of the hole in the grid is the crucial factor. The number of holes must not restrict normal flight activity at the entrance to the hive. All these factors must be considered when choosing your pollen trap.

Pollen traps are either front- or bottom-mounted. While front-mounted traps are certainly easy to install and remove, bottom-mounted traps are perhaps more efficient and effective. These traps can be housed in a standard hive body and they have a meshed screen bottom to aid ventilation and a drawer for collecting the pollen which slides out at the rear of the hive for easy access.

At the height of the flow, pollen should be collected on a daily basis and the trap should be removed as the pollen flow dwindles.

Above: *Pollen grains can be used as a health supplement.*

Storing the pollen

Freshly trapped pollen is perishable and must receive immediate attention to avoid loss. It can be dried, frozen or mixed with other materials and stored.

If you decide to dry the pollen, it is a good idea to freeze it overnight to kill off any wax moths or wax moth eggs that may be in it. Never use any pesticides as this will contaminate the pollen. When first collected, pollen has a moisture content of between 7 and 21 per cent, and this needs to be removed to prevent deterioration. You can purchase special pollen driers which are heated by an electric element, but when starting out, drying normally in the air will be sufficient for marketing.

To dry the pollen, spread it out on a flat, porous surface at a depth of about half an inch (12.7 mm) in an enclosed, ventilated room. A greenhouse is an ideal place to air-dry pollen. More rapid drying can be accomplished in ovens where a low temperature (37°C maximum) can be maintained and a vent provided for the moisture-laden air to escape. Pollen needs to be dried to the point where pellets will no longer stick to one another when squeezed. Dried pollen can be placed in an airtight glass or metal container and stored in a cool, dry place. Fresh pollen can also be placed in paper bags and stored in a freezer. It can be kept frozen until it can be dried.

Blending pollen with expeller-type soy bean flour is possible in equal parts by volume. Store this mixture in sealed containers in a cool, dry location. If you are blending, then you need to make sure that the pollen pellets and flour are pulverized and mixed thoroughly.

Cleaning pollen

Once the pollen has been dried thoroughly, you will need to clean it of any debris. The amount of debris will depend entirely on the efficiency of the pollen trap that you have used. Pollen can be cleaned by passing through various sieves of different calibres or differently-sized screens, so that the dust is collected in a box below the lowest screen.

Feeding pollen patties or substitutes

Bees need more than just carbohydrates from honey or sugar syrup to survive. They also need protein which usually comes in the form of pollen, especially when raising brood. You may find you never need to feed pollen or pollen substitutes as the bees will typically store enough for their own use, managing their supplies fairly well. However, there are times when additional pollen can be an advantage.

- Early spring build-up so you can make early splits.
- Build-up in preparation for pollination.
- To force building in preparation for a strong nectar flow.
- To encourage early drone rearing for preparation for raising early queens.
- To maintain drone and brood rearing through a strong dearth.

Below: *Bees feeding on a pollen patty.*

Pollen substitute can be fed when the weather is warm enough for the bees to fly out of the hive. Place the substitute in a simple container, such as a bucket, lay it on its side and allow the bees to take what they need. They will flock to it in their hundreds, but it does not tend to induce fighting amongst them.

You can also feed pollen patties which are placed directly inside the hive. These can either be purchased ready-made or you can have a go at mixing your own.

CHAPTER FIVE

HIVE PROBLEMS, PESTS AND DISEASES

Honey bees can be infected with various pests and diseases. Some are more harmful than others, but it is important for the beekeeper to be able to recognize problems and respond accordingly. Bees have two distinct life forms – brood and adult – and most of the diseases mentioned in this section are specific to one stage or the other. The most virulent diseases are those that affect the brood, such as American foulbrood and European foulbrood. Other brood diseases include chalkbrood, which is a fungal disease, and sacbrood which is caused by a virus.

Non-infectious disorders

This section is very important to beginner beekeepers, to allow them to recognize when a problem is not caused by an infectious disease. All brood and adult stages of queens, workers and drones may exhibit symptoms similar to those caused by disease, but due to other causes. Non-infectious disorders can result from neglect, overheating, chilling, poisoning from plants or pesticides or simply from queen failure.

NEGLECTED BROOD

Brood starvation

If there is a sudden dearth of pollen and nectar, the larvae and pupae may be removed or eaten by the adult worker bees. You will notice straight away that there is little or no brood, and very little honey or pollen present. There may be a few dead adult bees inside the hive, possibly with their heads still facing into the cells. This situation is called 'brood starvation'.

Treatment: Feed sugar syrup and protein supplement. Add frames to the hive containing mature brood, young bees and honey and pollen frames, first making sure they are disease-free.

Chilled brood

This will often happen in the spring when the queen has started laying. Sometimes she over-expands the brood area so there are not enough adult bees to tend the brood. In an unexpected cold spell, the bees will congregate to warm the central area of the brood, leaving the outer edges unprotected and they become chilled.

Chilled brood can also occur following an inspection in cold weather. If the frames are left exposed outside the hive for extended periods you risk losing your brood.

Chilled larvae and pupae will be yellow, which is tinged with black on their margins or a dull white with black or brown patches. The remains are pasty or watery.

Treatment: Take out any excess supers and feed sugar syrup if required.

Overheated brood

Overheating happens when the bees have been unable to control the temperature and humidity within the brood nest area.

There will be larvae hanging out of the top of their cells that are brown with a watery consistency. Pupae have a black, almost greasy appearance.

Newly emerged adult bees will be wingless, and cappings on brood cells may appear to have melted. Adult bees will become sticky, dark in colour, and run about noisily, desperately fanning their wings. There may be dead adults lying on the floor in the area between the frames.

Treatment: Make sure your hive has plenty of shade and ventilation and the bees have access to water. If the colony has been badly affected you may need to feed them with water or a dilute water/ syrup mixture. Remove any excess supers and dead adult bees. Add frames of mature brood and young bees.

OTHER PLANTS POISONOUS TO BEES

- California buckeye (*Aesculus californica*) – affects young brood
- Yellow jessamine (*Gelsemium sempervirens*) – affects larvae, pupae and young adults
- Loco plants (*Astragalus spp.*) – affects pupae and adults
- False hellebore (*Veratrum californicum*) – affects adults

PLANT POISONING

Not all plants are suitable for honey bees. Some produce nectar and pollen which are toxic to both adult and brood stages. Unfortunately, the bee is unable to identify which flower is poisonous as it is still attracted to the bright coloured foliage. An example of a poisonous plant is the Darling Pea (*Swainsona galegifolia*), which, if it is within a radius of hive that the foragers can reach, may cause heavy losses. Symptoms can be confused with a virus infection called 'sacbrood' disease.

Treatment: Move hives from the area and feed with sugar syrup if necessary.

Purple brood

Purple brood is a condition linked to areas abounding with the plant *Cyrilla racemiflora*, southern leatherwood or summer titi. Either the nectar and/or pollen of this plant is responsible for killing the brood and turning it a deep purple colour. Larvae, pupae and even newly emerged bees can be affected.

Ensure your apiary is not sited in an area where summer titi is in abundance. In addition, feed bees sugar syrup which helps to dilute the effects of summer titi on bee colonies.

POISONING FROM PESTICIDES

Signs of pesticide poisoning:

- Most or all of the hives in the apiary are affected.
- Adult bees usually die within a few days of each other.

Above: *Check the behaviour of the queen, the workers and the brood to make sure there are no problems occurring with the queen.*

- Dead adults normally have their wings unhooked, tongues fully extended and their hind legs stretched out behind them.
- In severe cases, dead adults will be present inside the hive.

Also look out for any abnormal behaviour. Sometimes adult bees may move very slowly. The brood may be dead. Queen failure or supersedure may take place within 30 days.

Treatment: Move hives away from the sprayed area, remove excess supers and feed the colonies inside the hives with a dilute sugar/water syrup. Be ready to manage the hives for queen failure or supersedure problems. Make sure you do not requeen until you are positive there is none of the chemical left in the hive. You will need to replace any contaminated frames which must be thoroughly cleaned before reuse.

QUEEN BEE PROBLEMS

Queen present

If you see any of the following it indicates you have a problem with your queen in that particular hive:

- The brood area reduces in size and has a scattered appearance, with brood of mixed ages in the same area.
- Excess pollen may be present in the brood area.
- Bee population numbers decrease and large numbers of drones may be reared in worker cells.
- Drone brood may not be fed sufficiently and could die.
- A small number of supersedure queen cells may be present, usually positioned on the upper third of combs in the brood nest area.

Treatment

Find and remove the original queen, queen cells and any emerged virgin queens. Requeen the colony with a mated queen or queen cell, or unite the colony with a strong colony using a sheet of newspaper between the two boxes. This will prevent fighting between adult bees and assist in the uniting process.

Queen absent

If there is no queen, the hive is termed 'queenless' and you will notice:

- There may be older stages of brood but not young.
- Pollen stores will have built up in the brood nest area.
- The hive bees emit a loud buzzing sound when the hive is initially opened.

Emergency queen cells

The brood, particularly drones, will be neglected and die. After a few weeks a small number of laying worker bees will be present. Laying workers are worker bees whose ovaries have developed slightly and are therefore able to lay a small number of unfertilized eggs. Signs of laying workers are:

- Too many eggs in each cell.
- Eggs laid around the side walls of the cell and not squarely in the bottom – these eggs develop into fully functional, but small, drones.

Above: *Spotty brood pattern with a multitude of empty and uncared-for brood cells.*

Treatment: Unite the affected colony with a strong, healthy one, using the newspaper method (see page 99). It is a difficult process, as the new queen is very often rejected.

Spotty brood

Spotty brood pattern is a sign of a bad queen. There are many empty cells where the queen has not laid eggs or diseased larvae have been removed by the bees. The most common cause is an old or failing queen.

A failing queen may also lay drone brood (or upright cells) interspersed with worker brood. Spotty brood can be a symptom of diseases like American foulbrood, European foulbrood, queenlessness, varroa mites, inbreeding or just a failing queen. It is always advisable to check for disease or other disorders when you see a spotty brood and requeen as necessary.

Brood diseases

AMERICAN FOULBROOD

American foulbrood (AFB) is by far the most virulent brood disease affecting honey bees. It is caused by the spore-forming bacteria *Paenibacillus* larvae. The spores of AFB can remain active for long periods of time which makes it difficult to get it under control. It attacks older larvae and young pupae, but does not affect adult bees.

A colony that has been infected with AFB will have a scattered and irregular pattern of both capped and uncapped cells. The infected cells will be discoloured and sunken, rather than the usual, slightly domed shape. The caps may also be punctured or perforated. As the cells die off, they turn from white to dark brown and death usually occurs in the final days of the larval stage or the first few days of the pupal stage. After death, additional bacterial spores will form, thereby increasing the spread of the disease.

For up to three weeks after death, the affected cells will have a glue-like consistency. If death occurs during the pupal stage, the pupae will go through the same change in colour and consistency as the larvae. In addition, the pupal tongue will stick up from the remains facing towards the top wall of the cell. This, along with the distinctive sour odour, is one of the most characteristic symptoms of AFB.

To prevent AFB

- Never allow bees access to honey, cappings and hive scrapings.
- Never interchange combs between infected and healthy hives.
- Never use second-hand equipment that hasn't been thoroughly cleaned.
- Never allow a hive to become neglected.
- Always inspect all brood combs each spring for the presence of AFB. Early detection increases the chance of saving the hive from destruction.
- Always sterilize your hive tools and wash the smoker, your hands or gloves after working on a hive that you suspect may be infected.
- Never remove a hive infected with AFB from an apiary until the infection has been eradicated.

You can use an approved antibiotic – terramycin – for preventative treatment against AFB. This antibiotic will not kill the *Paenibacillus* spores, but will prevent or delay their growth when present in low concentrations.

Preventative treatments of terramycin are normally made in early spring, well in advance of the major nectar flow. It is advisable to repeat the treatment in autumn after the honey crop has been removed.

AFB cannot be transmitted to humans and it has no effect on honey for human consumption. Because it is so contagious, every beekeeper needs to know the symptoms and should be able to recognize AFB in its early stages. If you suspect your hive is contaminated, and need help in the diagnosis, contact your local apiary inspector.

EUROPEAN FOULBROOD

European foulbrood (EFB) is also known as European brood disease. It is caused by the bacterium *Melissococcus pluton*, and can cause extensive losses.

Below: *Look out for brood diseases when carrying out general checks on your hives.*

Look out for:

- Brood with a mottled, peppered appearance with healthy brood cells intermingled with dead or dying ones.
- Larvae are mostly affected in the unsealed, curled-up stages..
- Diseased larvae collapse and become dislodged from their normal position in the cell. Their colour changes from pearly white to yellow, and then to a yellowish brown.
- In some cases sealed brood is affected and the capped brood takes on a mottled appearance with scattered sunken and perforated cappings.
- The odour of infected brood varies from odourless to sour or even foul, depending on the amount of bacteria present.
- Outer combs of the brood nest may show signs of the disease earlier and may have a heavier infection than inner combs in the same colony.
- If you probe the dead brood with a matchstick it usually has a watery consistency, although the sealed brown pupae may have a slightly stringy consistency.
- Worker bees may remove and discard diseased larvae, making it difficult to identify the disease.

When looking for EFB, pay close attention to any combs containing unsealed brood. Shake or brush the bees from combs. Hold the comb so that light penetrates the base of the brood cells being examined. Examine each comb in a regular pattern.

Diagnosis and treatment of EFB

The only accurate diagnostic method is laboratory examination and the only antibiotic recommended for the treatment of EFB is oxytetracycline hydrochloride.

There are a number of ways to treat colonies for EFB. If the apiary has a history of the disease, then all your colonies should be treated. This is best done well before any anticipated honey flows, usually in early spring or late autumn.

The only recommended method of administering the antibiotic is dry feeding. Feed each full-sized hive with 1 g of solution mixed thoroughly in 100 g of castor sugar. If you purchase one that has already been mixed with sugar, this can be used direct from the container.

Wear a face mask and gloves and avoid getting dust on your skin, breathing it in, or ingesting it. Apply the dry mixture by sprinkling between brood nest frames. Queen excluders should be removed before treatment. Treatments should not be applied by dusting the face of combs with powder as this can cause additional stress to the colony when the larvae are exposed to concentrated antibiotic.

Alternative treatments

Although there may be times when antibiotic treatment is the only answer, the practice is becoming less popular because of the possibility of contaminated honey. If you would like to try alternative treatments instead, then you can consider any of the following:

- Requeen your hive on a regular basis. Try to always select disease-resistant breeding stock.
- Maintain hive hygiene. Replace two or more white combs of foundation in the brood nest each year.
- Move bees with care. Bees are far more likely to shows signs of EFB just after they have been moved.
- Maintain nutrition. If ample honey is stored, lack of nectar should not be a problem. A good supply of pollen with adequate protein levels and a well-balanced group of amino acids is essential to reduce nutritional imbalance and stress on the bees. A lack of good-quality pollen can be overcome by artificially feeding previously collected pollen or pollen substitutes.

Although EFB is a serious disease, with careful hive management it can be reduced.

CHALKBROOD

Chalkbrood is a fungal disease – *Ascophaera apis* – which is characterized by infected brood called 'mummies'. When the brood are removed from the comb they appear to be solid clumps, which resemble pieces of chalk. Larvae are most susceptible to infection at three to four days of age and infected larvae die within two days of cell capping.

There is no recommended chemical control for chalkbrood, but requeening a colony is often a very effective treatment. The bees themselves can help to keep the disease under control by quick removal of the mummies.

Chalkbrood often appears in a peak in the late spring or early summer, as the colony starts to expand and the brood outnumber the bees. This is because there are insufficient bees to maintain the temperature and control the build up of CO_2.

You need to be careful in differentiating between chalkbrood and mouldy pollen, but the latter is generally concentrated around the periphery of the brood nest and tends to be a different colour.

As there are no fungicides available to destroy chalkbrood, any heavily affected comb should be destroyed. Viable spores will still be present on the bees and possibly in the honey stores. In severe cases requeening from a disease-free colony is recommended. Making sure there are enough bees in the colony to be able to control the temperature and humidity in the hive will greatly reduce the risk. Some strains of bee are more resistant and queens from these should be selected as part of an integrated breeding policy.

Above: *The distinctive mummified larvae that characterize an outbreak of chalkbrood.*

SACBROOD

Sacbrood is caused by a virus which attacks the larvae of the brood, killing them. Sacbrood attack reduces the number of larvae, which will have the effect of reducing the adult population. Although sacbrood does not wipe out the whole colony, it will weaken it considerably and make it susceptible to other pests and diseases.

The first symptoms show a change in colour of the larvae in open or closed cells of the comb. The diseased larvae change from their normal pearly white to a greyish-white gradually darkening to a yellowish colour with a black head. As the larvae begin to dry up, they become dark brown and start to shrivel. At a later stage of the attack, some of the larvae develop into a sac-like form which contains liquid, hence the name 'sacbrood'.

Check the hive for sacbrood once every two weeks and if you find any trace of the virus, take the following actions:

- If the attack affects more than 20 per cent of the comb, remove the comb and burn it, or melt it in a wax extractor.
- If more than 20 per cent of the brood is infected, all adult bees must be destroyed with a household aerosol spray. To ensure all bees are dead, close the hive and spray it again after 24 hours.
- Separate the comb wax from its frame and burn or melt the wax.
- Fumigate the hive's frame with formalin. This can be done by putting the entire frame in a plastic bag together with a small piece of cotton wool soaked with 40 per cent solution. Close the end of the plastic bag and leave for 24 hours.
- Isolate the hive box and sanitize before reusing. Either fumigate it with formalin, or submerge it in 2 per cent Dettol solution.

Managing an infected colony

If 5 to 20 per cent of the brood is infected, the colony usually recovers when the queen is replaced by a healthy, fertilized queen. Sometimes the colony can recover by itself by producing a new queen. If you do introduce a new queen, feed sugar syrup to speed up the development of the bee population. Where less than 5 per cent of the brood is infected, combine the colonies from two hives. Sugar syrup as an additional food should help the colony recover rapidly.

STONEBROOD

Stonebrood is caused by a fungus belonging to the genus *Aspergillus*. It attacks the brood and transforms the larva into a hard, stone-coloured object which is found lying in open cells. It is extremely rare and is presently of minor importance, but should not be totally ignored by beekeepers.

BALDBROOD

This is not a disease but the result of the wax moth (*Galleria mellonella*) larvae chewing through the brood cappings or tunnelling below the surface of the comb. The larvae perforate the cappings, which are then chewed down by the worker bees. These uncapped larvae will usually emerge as fully developed adults, although a few malformed adults may result from contaminants which have come into contact with the developing larvae.

Another cause of baldbrood is generic, where the worker bees do not cap the cells properly. Make sure you examine the brood frames and floor debris, especially in spring. There is no treatment for baldbrood other than requeening, or treating for wax moth.

Adult bee diseases

NOSEMA

Nosema disease is caused by a spore forming protozoan (*Nosema apis*) that invades the digestive tracts of honey bee workers, queens and drones.

When queens become infected with nosema, egg production and lifespan are greatly reduced which leads to supersedure. Infected workers, unlike healthy workers, may defecate inside the hive. Diseased colonies usually have increased winter losses and a much smaller honey production. Symptoms of the disease are not clear and sometimes, even at high levels of infestation, are difficult to detect. Things to look out for are unhooked wings, distended abdomens and what has been described as disoriented behaviour. Because the symptoms are so hard to spot, very often the majority of the bees in a colony are infected before nosema has been diagnosed. If you are suspicious that your bees may be infected, it is probably worth sending off a sample for analysis. The only positive way of identifying nosema is through the dissection of an adult bee. The hind gut and digestive tract of diseased bees are chalky white or milky white. In healthy bees, the gut will be amber or translucent. Also, the gut of an infected bee will probably be swollen and the individual circular constrictions are not easily visible.

Below: *Dead adult bees in brood cells.*

Controlling nosema

The best defence of all is to make sure that your colony is really strong before it goes into the winter dormancy. Make sure they have plenty of honey in the right position and a young, vigorous queen. Chemicals that have proved effective in treating nosema are Fumidil-B® or Nosem-X™ (Fumagillin). Fumagillin does not affect spores of the nosema parasite, so treatment with this drug will not completely eliminate the disease from the colony.

The best time for treatment is late autumn, when brood rearing normally declines. Feed about 3.5 litres of heavy sugar syrup (two parts sugar to one part water) containing the Fumagillin. The syrup should be stored where the last brood emerges and used as the colony's first winter feed. This will help prevent the initial build-up of any infection from winter confinement.

After the initial treatment, colonies should receive a minimum of 3.5 litres of sugar syrup containing 75 to 100 mg of Fumagillin in the autumn. Packages of newly installed bees in the spring should receive similar treatments. If your equipment has been infected by nosema, it can be decontaminated by use of heat (49°C for 24 hours). The temperature must not exceed this figure or the combs might melt. The technique must be used only on empty combs.

Warning: No medication should be fed to colonies when there is a danger of contaminating the honey crop. Be sure to stop all feeding of Fumagillin at least four weeks before the onset of the main surplus honey flow.

DYSENTERY

Dysentery in honey bees is not so much a disease but a symptom to show you that there is something wrong. It can develop as a result of other ailments such as nosema, or through feeding on fermenting honey. This can also occur if the bees are not able to perform their 'cleansing flights' to void their bowels due to a prolonged spell of cold weather. Dysentery problems usually get worse if there are periods of more than two or three weeks with temperatures below 10°C. The bee's gut will become engorged with faeces and the bee is forced to defecate within the hive. When enough bees do this the hive population rapidly collapses and death of the colony results.

Recognizing dysentery

The tell tale signs are soiled frames and combs and greatly increased faecal spotting on and around the hive's entrance. A bad case of dysentery can mean the entire front of the hive is covered with spotting. There may also be dead bees lying outside the hive entrance.

How to manage dysentery

Take a close look at the front, entrances, combs, frames and floor debris, especially in winter and spring months. Check all hives regularly for any signs of disease and make sure they are fed appropriately prior to and during the winter months. Any colonies that die out should be examined and sealed to prevent robbing and spread of any diseases present.

Prevention is definitely the best method of controlling dysentery. Good husbandry and apiary management contributes greatly to the overall health and behaviour of a colony, thereby avoiding the conditions in which dysentery can occur.

ACUTE BEE PARALYSIS VIRUS

Acute bee paralysis virus affects mainly the honey bee, but has also been found in bumblebees. It spreads by way of salivary gland secretions of adult bees and in food stores to which these secretions are added. In Europe and North America, ABPV has been shown to kill adult bees and bee larvae in colonies infested with the mite *Varroa jacobsoni*. The mite damages bee tissues and, in so doing, may act as a carrier, releasing viral particles into the hemolymph (the bee's equivalent to blood).

Suspected causes include pollen and nectar from plants such as buttercup, rhododendron, laurel and some species of basswood; deficient pollen during brood rearing in the early spring and consumption of stored fermented pollen.

Signs to look out for

- Affected bees tremble uncontrollably and are unable to fly.
- They lose their hair from their bodies and have a dark, shiny or greasy appearance.
- Large numbers of affected bees can be found at the colony entrance, crawling up the sides of the hive and blades of grass and tumbling to the ground.
- Healthy bees will often tug at the infected bees in an effort to drive them away from the hive.
- Close examination will often show that these bees have extended abdomens.
- Affected bees may also be found on top bars or frames next to the hive cover with their wings extended.

TRACHEAL MITES (ACARINE)

Tracheal mites (*Acarapis woodi*) are microscopic organisms that live in the trachea of the honey bee. These tiny parasites clog the breathing tubes of adult bees, blocking the oxygen flow and eventually killing them.

It is also known as Acarine disease. The reduced flow of oxygen affects the bee's ability to fly and often results in a large number of crawling or dead bees outside the hive. When an infected bee dies, the mites will abandon the carcass and move on to their next victim.

Identifying the mite

The tracheal mite is often very difficult to identify due to its microscopic proportions. It has an oval body which is widest between the second and third pair of legs. It is whitish in colour and has a shining, smooth cuticle. There are a few long hairs present on the body and legs, and it has a long, beak-like mouth for feeding on its host.

A bee infected with tracheal mites will have brown blotches on the trachea, with scabs or crust-like lesions. The trachea may also appear black due to the build-up of mites in their various stages of development. Feeding by the mites not only damages the trachea, but it can also affect the bee's flight muscles, causing them to be atrophied.

Treatment of the mite

Although there is no real way to prevent tracheal mites, it is possible to suppress the mite population. In the USA, menthol is officially registered for the control of tracheal mites in overwintering colonies. The menthol comes in pellet form and is applied to the hive by placing a packet on top of the top bars. Menthol works best when the outside temperature is approximately 21°C, and should be left in position for four to six weeks. Make sure you remove the packet before the colony beds down for winter.

Another method of treating the mite, which doesn't involve the use of any chemicals, is to use grease patties. Once it is placed inside the hive, the bees become greasy after coming into contact with the patty, which seems to interfere with the transfer of mites from one bee to another.

Below: *Varroa mites on a larval bee carcass.*

VARROA MITES

Varroa mites are small tick-like parasites that are native to Asia, using the Asian honey bee (*Apis cerana*) as their host. However, with the shipping of bees worldwide, there are only a few places left that do not have the mite. The adult mite is about the size of a pinhead and, unlike many of the other mites, is visible with the naked eye. The adult varroa mite lives on the adult honey bee, piercing its exoskeleton where it is relatively soft. This is usually between the segments of the abdomen where it feeds on the hemolymph. Young, developing varroa mites feed on the developing bee pupa causing a devastating effect on the colony.

Above: *An adult bee with a heavy infestation of varroa mites.*

The mites reproduce with the female entering the brood cells with the developing larva just before the cells are capped. Each cell can contain a number of mites, and they seem to prefer the drone cells rather than those containing workers. The mite starts to lay her eggs in a ten-day cycle, laying her brood on the bee larva. These eggs hatch between five and eight days later, with the males developing two days earlier than the females. These newly developed mites will mate in the cell. Once fully developed, the female mites emerge from the cell with the adult bee and survive by sucking the bee's blood. The males and any immature mites will stay behind in the cell and die. Because the mite prefers the drone cells, it is important to monitor the quantity of drone cells you have in a hive each time you carry out an inspection.

On adult bees the mites look like small, crab-shaped, red to dark brown blobs about 1.5 mm wide and 1.1 mm long. It is easy to miss them because they blend in very well with the overall colour of the bee, so you will need to look very carefully. If you suspect you have varroa mite, it is a good idea to uncap a drone brood, lift them out and give them a closer inspection.

Below: *Applying oxalic acid in water solution is one solution to treating an infestation of varroa in the hive.*

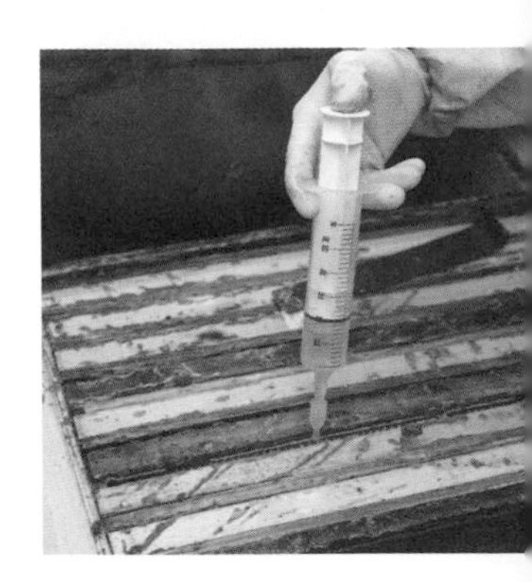

Controlling the varroa mite

Controlling varroa mite is a challenge.

Proprietary chemical treatments that are currently available include Apivar and Apistan among many others – check with your local organization for their preferred current options. They are all quick and easy to use so long as your mites are not chemical-resistant. There are also organic methods of treating varroa mite using natural products.

For example:

- Using essential oils, especially lemon, mint and thyme.
- A spray application of sugar esters (Sucrocide).
- Oxalic acid either used in a trickling or vapour method.
- Formic acid as a vapour or on pads of cotton wool.
- Food grade mineral oil used as a vapour or a direct application on paper or cord.

Some people prefer to use a physical or mechanical method which does not involve the use of any chemicals. These are intended to reduce the mite population to a manageable level, but will not eliminate the mites completely.

- Screened bottom board. When the mites fall off a bee, they have to climb back up to find a new host. If the beehive has a screened floor with mesh the right size, the mite will fall through but cannot return to the upper parts of the hive.
- Small cell foundation. By using foundation with smaller cells (about 0.5 mm smaller than standard ones), it limits the amount of space in each cell inhibiting the production of varroa mite larvae. Studies on this type of control have been inconclusive.
- Comb trapping. This is an advanced method of control involving the removal of capped brood from the hive where the varroa mites breed.

Behavioural methods

- Powdered sugar. Icing sugar or any other 'safe' powder with a grain size of between 5 and 15 micrometres can be sprinkled directly onto the bees. The powder inhibits the mite's ability to keep hold of the bee. It is also believed to increase the bee's grooming behaviour, which helps to dislodge a percentage of the mites.
- Freezing drone brood. Place a frame in the hive that is sized to encourage the queen to lay primarily drone brood. Once the cells are capped, remove the frame and put it in a freezer. This will kill the mites that are feeding off those bees. After freezing, the frame can be returned to the hive where the nurse bees will clean out the dead brood of bees and mites and the cycle will continue as normal.
- Hygienic behaviour. It is possible to breed good genetic traits into bees. This behaviour gene causes the bees to smell infected brood and remove them before the infestation spreads.
- Removing drone brood cells. Honey bees tend to place comb suitable for drone brood along the bottom and outer margins of the comb. If this is cut off at the later stage of development and then discarded, it

reduces the mite load on the colony. It also allows for inspection and counting of varroa on the brood.

PARASITIC MITE SYNDROME

Parasitic mite syndrome (PMS) is the name given to a range of abnormal brood symptoms associated with the presence of varroa and other mites in a colony. PMS is devastating to the entire colony because it affects both adults and brood and is usually associated with Colony Collapse Disorder. Symptoms can occur at any time of year, although most cases are reported from midsummer to autumn. The symptoms are not easy to detect because they can be confused with those of various other viral diseases, such as AFB, EFB and sacbrood. Listed below are the various symptoms that might arise, although please note that they may not all be present in any one colony at a given time:

Adult symptoms

- Varroa mites are present.
- Reduction in adult bee population.
- Evacuation of hive by crawling adult bees.
- Queen supersedure
- Tracheal mites may or may not be present.

Brood symptoms

- Varroa mites are present.
- Spotty brood pattern evident.
- Symptoms resembling EFB, AFB and sacbrood may be present.
- Individual larva ('C' shaped) may appear anywhere on the comb.
- Individual larva may appear twisted in the cell, stuck to the bottom of the cell and light brown in colour as in the early stages of AFB.
- You may see some scale formation but not as brittle as with AFB.
- There is no typical odour with this syndrome.

Control of PMS

There are no specific medications, but keeping up the general health of the bees and keeping down mite infestations is the best you can do to combat PMS.

TROPILAELAPS CLARAE

This mite is similar to the varroa mite, but can be identified by its elongated shape, as opposed to the crab shape of the varroa. Although this mite is not a problem outside of its natural Far East habitat, like varroa it could spread, and beekeepers are asked to look out for this new threat.

Apart from its shape, the other thing to look out for is an irregular, punctured brood pattern and malformed brood. The adult bees may have deformed wings.

The treatment is almost identical to varroa mite, but the mite does have one major weakness in that it cannot survive outside the cell for very long. This means when there is no brood, the hive will be clean of mites. The other major difference is that it can mate outside of the cells as well as inside.

SAMPLES FOR BEE DISEASE DIAGNOSIS

To the average beekeeper, recognizing bee diseases and their characteristics can be very confusing. It could cause a major problem if you were to wrongly diagnose, particularly when serious brood diseases are involved. For this reason it is advisable to contact your nearest laboratory, who will advise you where to send your sample and what further action may be required.

Contact your local beekeeping association for details.

WHAT TO SEND

Depending on which disease you suspect has caused the infestation, you can submit three types of samples – larvae (smears and mummies), adult bees or comb containing diseased brood.

Suspected disease	Sample
American foulbrood (AFB)	Larval smear (preferred) or comb sample containing diseased brood
European foulbrood (EFB)	Larval smear (preferred) or comb sample containing diseased brood
Chalkbrood	Mummies in or from comb
Sacbrood	Comb sample containing diseased brood
Nosema	Adult bees
Unknown disease	Contact an apiary officer immediately

Other pests and predators

WAX MOTH

Wax moths can be a major problem to bee hives if allowed to get out of hand, as they can destroy brood comb very quickly. A normal healthy hive will keep wax moth under control by ejecting the larvae, but weakened hives with small populations can quickly be overcome by infestations. They can destroy the brood comb until it is nothing more than a mass of web, ultimately destroying the hive.

There are two species of wax moth, the Lesser and the Greater, but it is the latter – *Galleria mellonella* – that causes the most damage. It is estimated to cause upwards of $5 million worth of damage a year in the United States alone.

Combating the wax moth

With a little care and careful hive management, the wax moth can be outwitted and the damage they do can be prevented.

Aways check the top entrances to your hives. If you have screening there should be no problem. If you leave a big hole in the inner cover and have a badly fitting roof, you are asking for trouble.

Make a lure to draw the moths away from your hives. Take a 2-litre plastic soft drinks bottle and drill a 25-mm hole just below the slope on the neck. Add a cup of water, a cup of sugar and half a cup of vinegar and finally the peel of one banana. Wait a few days for it to ferment, then tie it into a tree close to the hives. It will attract the wax moth. They will enter the hole and, unable to get out again, will drown in the liquid.

With minor infestations, pull out any larvae you can see and clean out all the webs. Freezing is also a very good way of killing both the larvae and the eggs, so storage in an outside, unheated shed during winter can be beneficial. Boxes should have a screen on both the top and bottom to not only prevent mouse damage, but to allow enough light to filter down as the wax moth prefers the dark.

There are preventative treatments on the market to treat boxes of brood comb if all else fails. Biological larvaecide is available for wax moth control. Use it just before storage or before the comb is placed in the hive. Prevention is better than cure, so just make sure the wax moth cannot get into your hive through any gaps and spaces.

Below: *A heavy infestation of wax moth – here as larvae – can devastate a hive.*

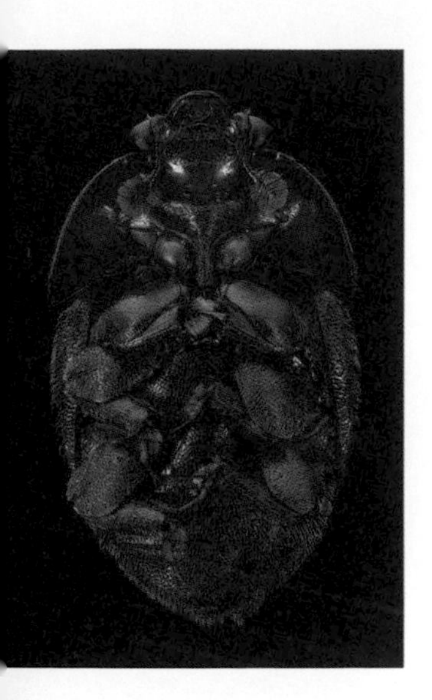

Above: *The small hive beetle can affect all parts of the hive.*

SMALL HIVE BEETLE

The small hive beetle (*Aethina tumida*) originated in Africa, and was first detected in Australia in 2002. Since then it has spread rapidly and is becoming a major problem in many parts of the world. It is considered to be a major pest as it can eat the brood, destroy the comb and quickly kill an entire colony of bees. Watch out for both the adult and the larvae, as both can be present in an active hive or in equipment which has been put into store for the winter. The adults have broad, almost flattened bodies that are about 5.7 mm long and 3.2 mm wide. They are very dark brown, almost black in colour. The larvae are white grubs with elongated bodies and rows of stubby spines on their backs. The eggs are laid in irregular masses all over the honeycombs.

To avoid confusion with wax moth larvae, the legs of the beetle larvae are larger, more pronounced and are situated near the head. The spines on the back also help to differentiate the two species. Beetle larvae do not spin webs or cocoons, but change to a pupae that develops generally in the debris at the bottom of a hive.

Although very small, the beetles are fast-moving and you can see them scurrying around when you first open the lid to your hive. The larvae burrow through the combs, eating the brood, pollen and honey on the way and leaving a repellent slime which acts as a deterrent to the bees, meaning they do not remove the larvae when doing their housekeeping. Any damaged combs will drip honey which then ferments.

Look for them in the bottom and corners of the brood box. You can usually get them to show themselves by leaving the super on an upturned lid for a few minutes. The beetles usually move on to the lid after they have been disturbed.

Or place some pieces of corrugated cardboard inside the hive, as the beetles will be attracted to them for shelter which makes them easier to spot.

It is thought that bees may carry the larvae on their bodies and as the beetles can fly up to 5 km they can travel to find a suitable host hive. Once a colony has been infested by this beetle, the bees usually abandon the hive, leaving a mess behind them because the beetles defecate in the honey and cause it to ferment.

Taking precautions

- Clean around the honey stores, making sure you do not leave any supers standing around before you carry out extraction. Do not leave any cappings exposed for long periods of time.
- Make sure your supers are not infested before placing them on top of strong colonies.

- Do not attempt to make splits or exchange combs between colonies if you suspect small hive beetle.
- Treat the soil in front of the hive with an approved insecticide.
- Treat colonies with an approved insecticide.

SQUIRRELS

Squirrels can chew and destroy combs. Make sure none of the hive entrances are big enough for a squirrel to get into, but large enough for your bees to come and go as they please.

MICE

Mice can be serious pests in a winter hive. Make sure to fix mouse guards on the entrance before winter. These usually consist of a metal strip with 9 mm holes drilled in it, which is large enough to allow the bees in and out, but not big enough for a mouse.

WASPS AND HORNETS

Wasps and hornets can be a nuisance in late summer/autumn by robbing honey from the combs. Restrict the hive entrance to make it easier for the bees to defend. Some beekeepers swear by wasp traps placed strategically near the hives.

WOODPECKERS

Woodpeckers can be particularly destructive in winter, drilling holes in the side of a hive. The holes are usually big enough for bees to crawl out of and the woodpecker will very often sit and pick them off as they emerge. You can cover the hives with chicken wire mesh or a layer of sacking; painting the hives white also seems to act as a deterrent as the woodpecker is not so keen to peck through the paint.

Below: *Attractive as they are, squirrels, mice and woodpeckers can all be destructive in the hive.*

GLOSSARY

Abscond – When bees leave a hive to locate a new nest area.

Alarm odour – A chemical (isopentyl acetate) released near the worker bee's sting, which alerts other bees to danger; also called alarm pheromone.

Alighting board – A small projection or platform at the entrance of the hive.

Antenna – One of two long segmented sensory filaments located on the head of the bee.

Apiary – A place where bees are kept.

Apiculture – The science and art of raising honey bees.

Bee blower – A gas or electrically driven blower used to blow bees from supers full of honey.

Bee bread – Pollen collected by bees and stored in wax cells, preserved with honey.

Bee escape – A device constructed to permit bees to pass one way, but prevent their return.

Bee space – The preferred distance between frames in a hive.

Bottom board – The floor of the beehive.

Brace comb – A bit of comb built between two combs to fasten them together.

Brood – The offspring of a queen.

Brood chamber – The part of the hive in which the brood is reared.

Burr comb – Comb built at odd angles or affixed to sides or the top of a hive box.

Candy plug – A fondant-type candy placed in one end of a queen cage to delay her release.

Capped brood – Immature bees whose cells have been sealed over with a wax cover by worker bees.

Cappings – The thin wax covering over honey.

Cell – The hexagonal compartment of a honeycomb.

Chalkbrood – A disease affecting bee larvae, caused by a fungus, *Ascosphaera apis.*

Cluster – Large group of bees hanging together.

Comb – Wax built into hexagonal cells used for rearing brood, and storage of pollen, nectar and honey.

Comb foundation – A structure consisting of thin sheets of beeswax with the cell bases of worker cells embossed on both sides.

Comb honey – Honey in wax combs, usually produced and sold as a separate unit.

Creamed honey – Honey that has been pasteurized and has undergone controlled granulation.

Cut comb honey – Comb honey cut into various sizes.

Dearth – A period of time when there is no available forage for bees.

Decoy hive – A hive placed to attract stray swarms.

Demaree – A method of swarm control.

Dequeen – To remove a queen from a colony.

Dextrose – One of the two main sugars found in honey.

Drawn combs – Combs with cells built out by honey bees from a sheet of foundation.

Drone – A male bee.

Drone congregating area (DCA) – A specific area to which drones fly waiting for virgin queens.

Drone laying queen – A queen that can only lay unfertilized eggs.

Dysentery – An abnormal condition of adult bees characterized by severe diarrhoea.

Entrance reducer – A notched wooden strip used to regulate the size of the bottom entrance.

Escape board – A board with one or more bee escapes in it. Used to remove bees from supers.

Extractor – Used to separate honey from comb.

Field bees – Worker bees usually over 21 days old who go out to forage.

Foulbrood – A severe bacterial disease of honey bees which is transmitted by spores.

Frame – Four pieces of wood forming a rectangle, designed to hold honeycomb.

Fume board – A device used to hold a volatile chemical to drive bees from supers.

Granulate – The process by which honey becomes solid or crystallizes.

Guard bees – Hive bees around the age of three weeks become guards and watch the hive entrance, only letting in bees that are part of the colony.

Hexagonal – Six-sided, the shape of cells in honeycomb.

Hive – A box in which bees establish a colony.

Honey flow – The peak of honey production, dependent on weather and food availability.

Instar – The developmental stages of a bee larva.

Italian bee – A strain of gentle honey bees.

Killer bee – The name associated with Africanized honey bees.

Larvae – Baby bees; the developmental stages of a bee from egg to pupae (cocoon).

Moisture level – A major difference between nectar (typically 20 per cent to 40 per cent water) and honey (less than 18 per cent water).

Nectar – Sugar solution provided by plants and collected by bees.

Newspaper method – A technique to join together, or unite, two separate colonies by providing a temporary barrier made out of sheets of newspaper.

Nosema disease – A honey bee disease caused by a protozoan.

Nuc, nuclei, nucleus – A small colony of bees often used in queen rearing.

Nurse bee – A bee at the life stage where she feeds and cares for developing bee larvae.

Pollen – The male reproductive cells of a plant, which provide proteins and nutrients for bees.

Propolis – A sticky plant compound collected by bees and used as a sealant in the hive.

Queen bee – The mother of all bees in a colony.

Queen substance – A pheromone (chemical scent) secreted by the queen which ensures the social cohesion of a hive.

Queen right – A colony with a healthy queen in residence.

Re queen – The practice of replacing the queen in a beehive every year.

Robber bees – Bees which sneak into weak or dying hives to steal honey or wax.

Royal jelly – A rich nutritive substance fed to the queen or to a bee less than three days old.

Sacbrood – A brood disease of bees.

Solitary bees – Bees which do not live in groups.

Split – To divide a colony for the purpose of increasing the number of hives.

Sugar syrup – Feed for bees.

Super – A segment of a hive, a box.

Supersedure – Rearing a new queen to replace the mother queen in the same hive.

Swarm – A collection of bees.

Top bar – The top part of a frame.

Tracheal mite – A mite which causes weakness or death in infected honey bees.

Uncapping – The removal of the wax cover of honey cells prior to extraction.

Varroa mite – An external mite which causes weakness or death in infected bees.

Virgin queen – An unmated queen bee.

Wax – Produced by wax glands on the 'belly' of a bee and used to construct the combs of a beehive.

Wax scale – A hardened piece of beeswax.

Wax moth – A pest of weak hives or stored hive boxes.

Worker – Most of the bees in a hive; they're all females except for drones.

INDEX